TRANSACTIONS

OF THE

AMERICAN PHILOSOPHICAL SOCIETY

HELD AT PHILADELPHIA

FOR PROMOTING USEFUL KNOWLEDGE

NEW SERIES—VOLUME 49, PART 5
1959

THE ANATOMY OF *CALLIMICO GOELDII* (THOMAS)

A Primitive American Primate

W. C. OSMAN HILL, M.D.

Prosector, Zoological Society of London

THE AMERICAN PHILOSOPHICAL SOCIETY

INDEPENDENCE SQUARE

PHILADELPHIA 6

SEPTEMBER, 1959

Library of Congress Catalog
Card No. 59–13421

ACKNOWLEDGMENTS

First and foremost my deepest gratitude is accorded to Dr. J. Tee-Van of the New York Zoological Society for his generosity in forwarding to me, on its death, the specimen of *Callimico goeldii* upon which the major part of this account is based, and to Dr. Leonard Goss, the Society's veterinary officer, for his ready cooperation in satisfying my suggestions and his prompt attention to the dispatch. I also wish to acknowledge the assistance of the Staff of the Mammal Room at the British Museum (Natural History), especially Mr. R. W. Hayman for permitting me to retain, on loan, for long periods, skulls of *Callimico* and several hapalid Primates. To Dr. G. E. Erikson of Harvard University I am indebted for certain information and for the photographs in figs. 2 and 3.

I wish also to acknowledge the facilities incident to my work in the Prosectorium of the Zoological Society of London, where the earlier part of the work was done including especially the technical staff of that Department as well as the Society's librarian, Mr. G. B. Stratton, and his assistants who, as always, have helped me in many ways especially in tracing obscure references. The later part of the work was done in the Anatomy Department, Emory University, Georgia, and for facilities there I have to thank the Chairman, Dr. Geoffrey H. Bourne. My thanks are due to the University librarians of Emory and to Miss Mildred Jordan, librarian of the A. W. Calhoun Medical Library at Emory, who have spared no pains in obtaining references for me, including some I was unable to obtain in London; also to Mrs. Mary Brandes for lettering of figures.

I am deeply grateful to Dr. G. W. Corner for his interest and advice.

KEY TO ABBREVIATIONS IN TEXT FIGURES

A.A.I.	Art. auditiva interna.	Ao.	Aorta.
A.A.P.	Art. auricularis posterior.	A. oe	Art. oesophagea.
A.Ax.	Art. axillaris.	A. Perf.	Art. perforantes.
A.B.	Art. basilaris.	A. Phr.	Art. phrenica inferior.
A.B.D.	Anterior belly of digastricus.	A.P.L.	Abductor pollicis longus.
Ab. H.	Abductor hallucis.	A.P.P.	Art. pancreatica propria.
A.B.P.	Art. brachialis profundus.	A. Pud.	Art. pudenda.
Ab. P.	Abductor pollicis.	A.R.Co.	Art. collateralis radialis.
A.Br.	Art. bronchiales.	A. Ren. D.	Art. renalis dextra.
A.B.S.	Art. brachialis superficialis.	Ary.	Arytenoideus muscle.
A.C.	Art. carotis.	A. Sc.	Atlanto-scapularis.
A.C.A.	Art. cerebralis anterior.	A.S.M.	Art. sacra mediana.
A.C.C.	Accessorius muscle.	A.Smd	Art. submandibularis.
A.Cer.Asc.	Art. cervicalis ascendens.	A.S.T.	Art. temporalis superficialis.
A.Cer. S.	Art. cerebellaris superior.	A.S. Th.	Art. thyroidea superior.
A.C.I.	Art. carotis interna.	A. Subsc.	Art. subscapularis.
A. Circ. A.	Art. circumflexa anterior.	A. Subcl.	Art. subclavia.
A. Circ. Med.	Art. circumflexa femoralis media.	A. Th. L.	Art. thoracalis longa.
A. Circ. P.	Art. circumflexa posterior.	A.U.	Art. ulnaris.
A. Comm. P.	Art. communicans posterior.	Au. P.	Auricularis posterior.
A.C.P.	Art. cerebralis posterior.	A.U.R.P.	Art. ulnaris recurrens posterior.
Ac. Th.	Art. thoraco-acromialis.	A.V.	Art. vertebralis.
Add.	Adductor mass.	B.B.	Biceps brachii.
Add. H.	Adductor hallucis.	B.C.	Bulbo-cavernosus.
A.D.M.	Abductor digiti minimi.	B.F.	Biceps femoris.
A.D.Q.	Abductor digiti quinti.	B.R.	Brachio-radialis.
A.Ep. Pr.	Art. epigastrica profunda.	Br.	Brachialis.
A. Fem.	Art. femoralis.	C. 7.	Cervical branch of seventh nerve.
A. fem. prof.	Art. femoralis profunda.	C.A.L.	Crico-arytenoideus lateralis.
A.G.E.D.	Art. gastro-epiploica dextra.	C.A.P.	Crico-arytenoideus posterior.
A.Gl. Inf.	Art. glutea inferior.	C.B.D.	Common bile-duct.
A. Hyp.	Art. hypogastrica.	C.F.	Caudo-femoralis.
A.I.	Art. interossea.	C.I.T.	Caudal intertransverse muscles.
A. Il. E.	Art. iliaca externa.	Cl.	Clavicle.
A. Int.	Art. intercostales.	Con.	Contrahentes.
A. Int. 2.	Art. intercostalis secunda.	Con. inf.	Constrictor pharyngis inferior.
A.L.I.	Art. laryngea interna.	C.S.T.	Caudal head of semitendinosus.
A.M.A.	Art. mesenterica anterior.	C. Th.	Crico-thyroideus.
A.M.D.	Art. marginalis dextra.	D.	Deltoideus.
A.O.	Art. occipitalis.	D.D.	Ductus deferens.
A. Ob.	Art. obturatoria.	D.I.	Dorsal interosseous muscle.

3

D.J.F.	Duodeno-jejunal flexure.	N. Cau. lat.	N. caudalis lateralis.
E.C.R.B.	Extensor carpi radialis brevis.	N. Cl. P.	NN. cluneum posteriores.
E.C.R.L.	Extensor carpi radialis longus.	N. Coc.	N. to cochlea.
Ec. G.	Ectoglutaeus.	N. Cut.	NN. cutanei.
E.C.U.	Extensor carpi ulnaris.	N. Cut. fem. post.	N. cutaneus femoralis posterior.
E.D.B.	Extensor digitorum brevis.	N. Cut. l.	N. cutaneus femoralis lateralis.
E.D.L.	Extensor digitorum longus.	N.D.I.	N. dentalis inferior.
E.D.M.	Extensor digiti minimi.	N. Fem.	N. femoralis.
E.H.L.	Extensor hallucis longus.	N. Gem.	N. to gemelli, etc.
E. H. Pr.	Extensor hallucis profundus.	N. Gf.	N. genito-femoralis.
Epig.	Epiglottis.	N. Gl. A.	N. glutaeus anterior.
E. Pr. I—III.	Extensores proprii digiti I—III.	N. Ham.	N. to hamstrings.
E. Pr. III, IV.	Extensores proprii dig. III & IV.	N.I.B.	N. intercosto-brachialis.
F.	Fat.	N.I.H.	N. ilio-hypogastricus.
Fb.	Brown fat.	N.I.I.	N. ilio-inguinalis
F.C.	Femoro-coccygeus.	N.I.P.	N. interosseus posterior.
F. Ca. L.	Flexor caudae lateralis.	N.L.A.	N. to levator ani.
F.C.U.	Flexor carpi ulnaris.	N.L.B.	N. buccalis longus.
F.D.B.	Flexor digitorum brevis.	N. Li.	N. lingualis.
F.D.P.	Flexor digitorum profundus.	N.L.S.	N. laryngeus superior.
F.D.Q.B.	Flexor brevis digiti quinti.	N. L. Sc.	N. to levator scapulae.
F.D.S.	Flexor digitorum sublimis.	N. Il.	N. to iliacus.
F.F.	Flexor fibularis.	N. I. Mj.	N. ischiadicus major.
F.f.	Floccular fossa.	N. L.A. T.	N. thoracalis anterior lateralis.
F.H.B.	Flexor hallucis brevis.	N.L.H.G.	N. to lateral head of gastrocnemius.
Fib.	Fibula.	N.L. Palp.	N. to levator palpebrae.
Fl.	Flexor tendons of tail.	N.L.R.	N. laryngeus recurrens.
F.L.R.	Flexor carpi radialis.	N.M.	N. medianus.
f.p.	Fissura prima cerebelli.	N.M.A. T.	N. thoracalis anterior medialis.
F.P.L.	Flexor pollicis longus.	N. Mc.	N. musculo-cutaneus.
f.pn.	Fissura postnodularis.	N. Md.	Mandibular division of n. trigeminus.
f.r.	Fissura rhinalis.	N.M.H.G.	N. to medial head of gastrocnemius.
Fr. Lam.	Frenal lamella.	N. Mx.	Maxillary division of n. trigeminus.
f.s.	Fissura secunda cerebelli.	N.N.C.B.	NN. ciliares breves.
f.sp.	Fissura suprapyramidalis.	N.N. Phar.	NN. pharyngeales.
F.T.	Flexor tibialis.	N.O.	N. opticus.
G. Ad.	Glandula adrenalis.	N. Ob.	N. obturatorius.
Ga.	Gastrocnemius.	N.O.I.	N. to inferior oblique.
Ga. L.	Gastrocnemius, lateral head.	N.O.M.	N. oculomotorius.
Ga. M.	Gastrocnemius, medial head.	N. Op.	Ophthalmic division of n. trigeminus.
G.B.U.	Glandula bulbo-urethralis.	N. Per Comm.	N. peronaeus communis.
G.C.	Ciliary ganglion.	N. Ph.	N. phrenicus.
G.G.	Gasserian ganglion.	N.P.L.	N. popliteus lateralis.
Gr.	Gracilis.	N. Pl. L.	N. plantaris lateralis.
G. St.	Ganglion stellatum.	N. Pud.	N. pudendus.
G. Sub.	Glandula submandibularis.	N. Py.	N. to pyriformis.
G. Thy.	Glandula thyroidea.	N. R.	N. radialis.
Hy.	Hyoid bone.	N.R.I.	N. to rectus inferior.
Hy. G.	Hyoglossus.	N.R.M.	N. to rectus medialis.
I.C.	Ischio-cavernosus.	N.R.S.	N. to rectus superior.
I. Ps.	Ilio-psoas.	N.S.	N. subcostalis.
I. Sp.	Infraspinatus.	N. Sa.	N. to saccule and posterior semicircular canal.
J.C.T.	Jugulo-cephalic trunk.		
L. Bf.	Lig. bifurcatum.	N.S.M.	NN. to sterno-mastoid.
L. Cap.	Longissimus capitis.	N. S. Sc.	N. suprascapularis.
L. Col.	Longus colli.	N. Subsc.	NN. subscapulares.
L.D.	Latissimus dorsi.	N. Sup.	N. to supinator.
L. Ep.	Latissimo-epicondyloideus.	N. t.	N. to trapezius.
Lg.	Lymph-glands.	N. Th. L.	N. thoracalis longus.
L.H.T.	Long head of triceps.	N.T.P.	N. tibialis posterior.
Lig. P.	Lig. patellae	N. Tib. A.	N. tibialis anterior.
L. Orb.	Lig. orbicularis radii.	N. Tib. P.	N. tibialis posterior.
L.P.S.	Levator palpebrae superioris.	N.U.	N. ulnaris.
L. Te.	Left testis.	N. Ut.	N. to utricle and lateral semicircular canal.
L. Tm.	Lig. talo-metatarsale.	N.x.	N. vagus.
Lum.	Lumbricales.	Ob. I.	Obliquus inferior (capitis).
M.	Masseter.	Ob. S.	Obliquus superior (capitis).
M.G.	Mesoglutaeus.	Occ.	Occipital slip of sterno-mastoid.
M.H.	Mylohyoideus.	Oe.	Oesophagus.
M. Tub.	Median tubercle of soft palate.	O.E.A.	Obliquus externus abdominis.
N. Ax.	N. axillaris.	O.H.	Omohyoideus.

O.I.	Obliquus inferior (bulbi).	S.H.	Sterno-hyoideus.
O. Mi.	Omentum minus.	S.H.T.	Short head of triceps.
O. Mj.	Omentum majus.	S.M.	Sterno-mastoideus.
O.S.	Obliquus superior (bulbi).	S.Me.	Semimembranosus.
O.7.	Occipital branch of VIIth nerve.	S.M.L.	Stylo-mandibular ligament.
P.A.	Pectoralis abdominis.	Sol.	Soleus.
P.Ac.	Peronaeus accessorius.	Spc.	Splenius colli.
Panc.	Pancreas.	Spl.	Spleen.
Pa. L.	Palmaris longus.	S.S.	Slip from flexor sublimis.
P.B.D.	Posterior belly of digastricus.	S.S.C.	Semispinalis capitis.
P. Br.	Peronaeus brevis.	S.T.	Semitendinosus.
P.C.	Pubo-coccygeus.	St. G.	Stylo-glossus.
P. Co.	Posterior cord of brachial plexus.	St. H.	Stylo-hyoideus.
P.D.	Parotid duct.	St. Ph.	Stylo-pharyngeus.
P.G.F.	Palato-glossal fold.	Subcl.	Subclavius.
P.I.	Plantar interrosseous muscle.	Subsc.	Subscapularis.
Pis.	Pisiform.	Sup.	Supinator muscle.
Pit. B.	Pars buccalis of pituitary.	S.V.	Serratus ventralis.
Pit. N.	Pars nervosa of pituitary.	Sy.	Sympathetic chain.
Pl.	Plantaris.	T. 7.	Cervical branch of n. facialis.
P. Lo.	Peronaeus longus.	T. Ac.	Tendo Achillis.
P.M.	Pectoralis major.	T.B.	Triceps brachii.
P. Min.	Pectoralis minor.	T.B. (L)	Lateral head of triceps.
Pop.	Popliteus.	Tem.	Temporalis.
Pop. V.	Popliteal vessels.	Ten.	Tenuissimus.
P.Q.	Pronator quadratus.	Tent.	Tentorium cerebelli.
Pro.	Prostate.	Th. Ary.	Thyro-arytenoideus.
P.R.T.	Pronator radii teres.	Th. H.	Thyro-hyoideus.
P.S.M.	Presemimembranosus.	Thym.	Thymus.
Pt. M.	Pterygoideus medialis.	T.L.F.	Truncus linguo-facialis.
Q.F.	Quadriceps femoris.	T. Mi.	Teres minor.
R.C.P. Mi.	Rectus capitis posterior minor.	T. Mj.	Teres major.
R.C.P. Mj.	Rectus capitis posterior major.	T.P.	Tibialis posterior.
R.C.V. Mi.	Rectus capitis ventralis minor.	Tr. M.	Trachelo-mastoideus.
R.C.V. Mj.	Rectus capitis ventralis major.	U.B.	Urinary bladder.
Rect.	Rectum.	Ur.	Ureter.
Rh.	Rhomboideus.	V.	Vessels.
R. Inf.	Rectus inferior (bulbi).	V. Az.	Vena azygos.
R.L.	Rectus lateralis (bulbi).	V.C.A.	Vena cava anterior.
R.M.	Rectus medialis (bulbi).	V.C.P.	Vena cava posterior.
R. Mot.	Radix motorius n. trigemini.	V.J.	Vena jugularis interna.
R.S.	Rectus superior (bulbi).	V.L.	Vastus lateralis.
R. Sen.	Radix sensorius n. trigemini.	V.M.	Vastus medialis.
Sc. M.	Scalenus medius.	V.S.	Vesicula seminalis.
Sc. P.	Scalenus posterior (dorsalis).	V.V.P.	Venae pulmonales.
Sc. V.	Scalenus ventralis.	Y.	Y-nerve of Appleton.
S.F.	Sinus frontalis.		

THE ANATOMY OF *CALLIMICO GOELDII* (THOMAS)

A Primitive American Primate

W. C. OSMAN HILL, M.D.

CONTENTS

INTRODUCTION

The interest in *Callimico*—apart from its extreme rarity—lies in its allegedly intermediate taxonomic status astride the commonly accepted definitions of the two extant families of platyrrhine Primates—the Hapalidae (= Callithricidae) and the Cebidae—possessing, as it does, the extremities of the former coupled with the dental formula of the latter. Thus far, however, it is known solely from the museum systematist's standpoint, namely on the basis of dried skins and a small number of skulls (mostly damaged). Pocock (1920), however, gave some brief data on the nose, ear, extremities, and male genitalia of a fresh specimen. The present writer, too, has published (1957) a *précis* of the principal findings reported *in extenso* in the present memoir.

The history of our knowledge of this interesting genus may be briefly recalled. Thomas (1904) gave the name *Midas goeldii* to an incomplete male specimen without skull, sent to him by Goeldi. The animal had lived for a short time in the zoological garden attached to the Pará Museum. Goeldi (1904) had considered it to be an aged example of Weddell's tamarin (*Tamarinus weddelli*) though recognizing some aberrations in pelage. In 1911 Miranda Ribeiro saw a second living example in the Pará menagerie. Unaware of, or else failing to recall, Thomas's description but recognizing, on the grounds of its appearance and behavior, its distinctness from known tamarins, Ribeiro (1912) described it as representing a new genus, giving it the name *Callimico snethlageri*. On its death, this specimen, a female, was sent, this time complete, to the British Museum, whereupon Thomas (1913) recognized its identity with his *M. goeldii*. He now concurred with Ribeiro in separating it generically, so that by the rules of nomenclature its name becomes *Callimico goeldii*. Ribeiro suggested his generic name because the animal showed characters intermediate between *Callicebus* and *Mico*.

Elliot (1913) in the first volume of his monograph (p. 224) placed the animal in the genus *Callithrix* as *C. goeldii* alongside the typical marmosets, but in an appendix to his final volume (p. 261) he had accepted Thomas' more recent findings and, on account of the dental formula, elevated *Callimico* to a subfamily of the Cebidae.

Thomas (1914: 345) again referred to *Callimico* on the receipt of a further juvenile specimen from the Pará menagerie, but, in this instance, the exact original provenance was known, namely the Río Xapury, an affluent of the Río Acre, Upper Río Purús. He recorded that "milk premolars" were still *in situ*, but the characteristic third molar of the new genus was present below the level of the alveolar bone. This specimen appears subsequently to have been returned to Pará.

The following year on the fourteenth of April, a living example was deposited at the Zoological Society of London by Lt. F. D. Walker, R.A.M.C., who had brought it from the Mu River. It survived only until April 20, 1915. Recorded on receipt as a Negro tamarin (*Midas ursulus*) now known as *Tamarin tamarin* or *Leontocebus tamarin*[1] its true identity was evidently established after death, for it is presumably on the basis of this individual that Pocock, then Superintendent of the Society's Gardens, made the aforementioned observations. I have endeavored for years to trace the fate of this specimen, without success except for the fact that, at the depositor's request, skin and skull were sent to the British Museum where they still remain.

The first wild killed specimens were two females obtained at Cerro Azul, Contamana, in northwest Peru by R. W. Hendee referred to by Thomas in 1928. In this paper Thomas noted that in all the four examples he had previously seen, the hairs on the throat and interramal region inclined medially to form an incipient median crest, pointing forwards, in contrast to all the marmosets where this region is smooth, with backwardly directed hairs. He had by now come to the same conclusion as Pocock, that *Callimico* was a primitive hapalid rather than a member of the Cebidae.

Except for some further comments by Ribeiro (1940) —chiefly relating to the skull—no further reference occurs in the literature until 1945, when Lima's monumental work on the Amazonian Primates appeared. In this is given a colored plate of *Callimico*, based upon a living animal in the Pará menagerie. Its skin is preserved in the Goeldi Museum together with two others, one of which (from the Xapury River) is believed to be the one mentioned by Thomas in 1914.

In the latter part of 1954 the New York Zoological Society acquired, through a dealer, a living *Callimico*. It died on January 21, 1955, and was most generously forwarded to the writer for anatomical study, the present accounting being based principally thereon.

Thereafter the Bronx Zoo received three other living specimens, a male and two females. After death these fell into the hands of my friend Dr. G. E. Erikson of Harvard Medical School who very kindly supplied me

---

[1] To maintain uniformity and avoid confusion the generic nomenclature used for the different groups of marmosets and tamarins is that of Hill, 1957. It should be mentioned, however, that in one respect this was already out of date at the time of publication and while the bulk of the present memoir had been written.

New information (*fide* Cabrera, 1956; Hershkovitz, 1957) demands the transfer of the generic name *Leontocebus* to the white-faced group of tamarins and therefore replaces *Tamarinus* as used by Hill, 1957. Cabrera (*loc. cit.*) suggests the new name *Leontideus* for *rosalia* and its allies (i.e., the *Leontocebus* of Hill, 1957).

with additional information based upon these animals and provided the photographs of the external genitalia of both sexes reproduced on figure 6.

MATERIAL

The specimen (No. PP77) was forwarded on condition that a post-mortem examination was carried out to ascertain the cause of death and also that the skin and skull should be returned for presentation to the American Museum of Natural History, in whose collection the genus was not then represented. It was, therefore, at my request, sent freeze-dried by air transport and arrived, in the circumstances, in excellent condition some days after its demise. I was thus enabled (1) to take measurements and make observations on the complete carcass, supplementing and adding much detail to Pocock's rather scanty remarks on the external anatomy, and (2) to obtain radiographs of the total animal before removing the skin. In preparing the skin, the hand and foot of the left side were removed at wrist and ankle, but on the right the extremities were skinned as far as the digits, so as to retain the short muscles. After skinning, the abdomen was opened and the principal organs examined *in situ* while fresh. The specimen was then given a red injection mass through the abdominal aorta and the carcass thereafter immersed in 10 per cent formalin for fixation, being afterwards transferred to 70 per cent alcohol for dissection.

All the material in the British Museum (Nat. Hist.) has also been examined (table 1).

TABLE 1

Registration No.	Sex	Remarks
0/2/22/1	♂	Type of *Midas goeldii* Thos. Skin only. Ex Pará Museum.
12/11/4/2	♀	Type of *Callimico snethlageri* Rib. Skin and skull. Ex Pará Museum.
39/608	♂	From Zoological Society of London, April, 1951, originally from Mu River, coll. F. D. Walker. Skin and skull (death book no. 385/15).
28/5/2/66	♀	From Cerro Azul, Contamana, Peru; coll. R. W. Hendee. Skin and skull.
28/5/2/67	♀	From Cerro Azul, Contamana, Peru; coll. R. W. Hendee. Skin and skull.

After the completion (and acceptance for publication) of the present monograph two additional specimens of *Callimico,* a male and a female, were received in excellent condition thanks to Mr. Ivan Sanderson of New York. These have not been completely dissected, but have enabled me to fill several important gaps in the original account, notably regarding the female genitalia, both internal and external, the mammary glands and the facial muscles. Table 2 has also been enlarged to accommodate the measurements of the new specimens.

GENERAL DESCRIPTION

As recorded by Thomas (1928) *Callimico* in general appearance strongly recalls a tamarin, but in the freshly dead cadaver (as in life, *vide* fig. 1) there are many specific features distinguishing it from all others. It has the same orthognathous face and dolichocephaly of such types as *Leontocebus* (*Leontideus*) *rosalia* and in its mandibular incisor-canine relationship it is normal, lacking the specialized incisor elongation of the true or "short-tusked" marmosets (*Hapale, Mico,* and *Cebuella*).

The *pelage* is characteristically soft, sleek, and silky; elongated and densely planted dorsally and on the extensor aspects of the limbs, as well as on the tail, but sparsely developed below and on the flexor surfaces. On the crown the hair is porrect; but, from a radiation on the nuchal region, a pair of upstanding plumes proceed vertically and craniad, forming a kind of pompador on the back of the head, projecting some distance beyond the tips of the porrect hairs of the forepart of the crown. The appearance recalls somewhat that of the cebid *Chiropotes* and was appropriately likened by Ribeiro (1912) to the head adornment of the Crowned crane (*Balearica regulorum*). The hairs are the longest on the body, measuring 26 mm. Hairs on the sides of the head radiate from a preauricular center, those directed downwards being much elongated, proceeding downwards and forwards, forming converging tufts near the chin, though leaving the chin itself comparatively bare (*c.f.* Thomas, 1928). The ears are completely hidden from view, but differ from those of the hapaline marmosets (though agreeing with such genera as *Mico, Marikina,* and *Oedipomidas*) in lacking intrinsic tufts or other aural adornments. A curious bare tract surrounds the attachment of the ear all around, the hairline ceasing abruptly like that over the mastoid region in Man.

PELAGE: COLOR

In color the pelage is predominantly black everywhere. Goeldi's original type showed some irregularly distributed white flecks which were at first attributed to age or injury (Thomas, 1904). Lima's account too mentions paler areas, notably a double stripe "tawny blond and white, slightly tinged with rufous" above the nape, and also on the loins and two or three annuli on the tail.

Lima's plate repeats these features and it is notable that he shows the pale tipped nuchal hairs as lying depressed over the nape, whereas in the New York example they were upstanding and antrorse, as de-

FIG. 1. *Callimico goeldii* ♂. Living specimen in the New York Zoological Society's Menagerie, Bronx Park, New York. (Courtesy of New York Zoological Society.)

scribed by Ribeiro (1912). There is no evidence of their being erectile during life and it is difficult to reconcile the two descriptions. Lima's account moreover is not, as might have been suspected, based on preserved skins, but on a living animal in the Pará Zoo.

The New York example is entirely black, except for a slight tawny ticking, due to the presence of that hue on the terminal 3 mm. of the elongated nuchal hairs and on most of those of the back, giving an effect as if they had been slightly singed. This surface wash is quite lacking on the hairs of the forehead and anterior part of the crown, and also on the tail, hands, and feet, which are undiluted black throughout.

Juveniles, according to Lima, are entirely brownish-black with only the tips of the hairs on the loins a lighter brown.

CUTANEOUS PIGMENTATION

The distribution of cutaneous pigment is of some interest. The face is deeply pigmented throughout, except for some lightening on the eyelids and orbital regions. The ears are pigmented only near their free margins, i.e. over a zone some 5 mm. wide, including the postero-superior part of the helix, the remainder being pink. Some pink also shows on the throat and pectoral region, but otherwise the under parts are dusky. Palms and soles are black, but on the palms the pigment is lessened in irregular patches (*a*) on the thenar eminence, (*b*) over the radial interdigital pad, and (*c*) over the proximal interphalangeal joints, where a pinkish tinge shows through the epidermis. No corresponding pallid areas were found on the sole. Pigment extends into the mouth, on to the gums, cheeks, hard palate, and frenal lamella, but not to the tongue (*c.f. Cebus* etc., Sonntag, 1921). The nictitating membrane and sclera are also black, as noted by Ribeiro (1940). Pigmentation in the genital zone is discussed below. General pigmentation of the body is lacking.

All the digits except the hallux are provided with claw-like nails as in all the true tamarins and marmosets, but they are relatively stouter, less recurved and less acutely pointed, differing little, therefore, from the intermediate condition found in such cebids as *Saimiri* (Dollman, 1937). Structurally they may be regarded therefore as hapalid claws as described and figured by Bruhns (1910) and Clark (1936).

MICROSCOPIC ANATOMY OF HAIR (FIGS. 2, 3)

Samples from various parts of the body were submitted to Mr. H. M. Appleyard of the Wool Industries Research Association, Leeds, for investigation and comparison with other hairs from various platyrrhine monkeys then being studied in another connection (Hill, Appleyard and Auber, 1959).

There is nothing in the structure of the hairs from *Callimico* fundamentally different from those of other Platyrrhini, whole mounts and longitudinal sections (stained with basic fuchsin and picro-indigo-carmine) failed to reveal the presence of medullary substance. Melanized granules are present, usually situated at the distal side of the cortical septa which protrude into the medulla. Cuticle is of medium thickness. On treatment with 17 per cent caustic potash flattened medullary alveoli seen in whole mount assume a spheroid shape, as in *Hapale*.

Photomicrographs of hairs of *Callimico* are compared with similar preparations from *Hapale jacchus* in figures 2 and 3.

SOMATOMETRY

The body weight of the New York specimen on receipt was 278 gms. Below (table 2) are given the linear dimensions compared with those of a male Negro tamarin (*Tamarin tamarin*) followed (table 3) by the

TABLE 2

SOMATOMETRY OF CALLIMICO (IN MILLIMETERS) COMPARED WITH THAT OF TAMARIN

	PP 77 ♂	PP 118 ♂	PP 119 ♀	T. tamarin ♂
1. Sitting height (vertex-rump)	215.0	250.0	245.0	233.0
1a. Total length (muzzle-rump)	220.0	252.0	240.0	237.0
2. Thoraco-abdominal height (symphysion-suprasternale)	146.8	154.0	151.5	150.0
3. Symphysion—nipple	—	130.0	130.0	124.5
4. Symphysion—omphalion	[1] 56.0	72.0	—	—
5. Bisacromial diameter	[1] 45.0	52.0	54.0	53.5
6. Bimammillary diameter	[1] —	34.5	41.0	31.0
7. Bitrochanteric diameter	[1] 37.8	42.0	47.5	36.4
8. Transv.-thoracic (nipple level)	[1] 42.5	44.8	41.0	38.5
9. Sagittal diameter of thorax	[1] 50.0	50.0	53.0	47.8
10. Circumference of thorax	[1]150.0	160.0	155.0	142.0
11. Length upper arm (caput-radiale)	[1] 49.7	58.8	53.3	48.0
12. Length forearm (radiale-stylion)	[1] 51.0	57.0	53.0	48.1
13. Length manus (styloid-tip of medius, excl. claw)	44.0	44.0	45.5	40.0
14. Length thumb	8.0	12.0	11.1	8.5
15. Breadth of manus (II-V)	12.8	15.5	13.7	14.8
15a. Max. br. manus	16.0	19.8	—	17.4
16. Length femur (top of trochanter-tibiale)	[1] 69.3	67.4	63.0	70.0
17. Length crus (tibiale—malleolare)	[1] 73.5	76.0	78.0	71.5
18. Tibiale—sole	[1] 82.0	78.0	85.0	72.2
19. Length pes (heel—tip of annularis)	68.6	76.0	69.2	67.3
20. Breadth pes (II-V + br. of joint of hallux)	15.8	20.0	20.0	15.4
20a. Breadth pes beyond hallux	13.0	18.5	15.2	14.0
21. Maximum length head (glabella—maximum occipital point)	55.0	55.0	46.5	46.0
22. Maximum breadth head	32.9	40.5	36.0	33.0
23. Auricular height of head	23.5	28.4	26.5	20.0
24. Nasion—inion	45.0	46.2	46.0	43.7
25. Biauricular	[1] 36.3	37.0	24.5	33.8
26. Circumference of head	[1]138.0	145.0	140.0	140.0
27. Sagittal arc (nasion-inion)	[1] 63.0	68.0	61.0	60.0
28. Transv. arc (tragion-tragion)	[1] 72.0	71.0	73.0	53.0
29. Gnathion-vertex	37.5	39.0	37.6	33.1
30. Nasion-gnathion	25.1	22.0	22.6	23.0
31. Nasion-cheilion	16.2	14.0	13.5	11.6
32. Bizygomatic breadth	36.0	34.0	33.6	34.5
33. Nasal ht.	10.1	10.0	10.0	8.5
34. Nasal br.	10.5	11.8	12.0	8.2
35. Br. nasal septum	6.2	7.3	7.9	7.0
36. Interocular breadth	7.0	10.0	10.0	9.1
37. Br. mouth	17.8	19.1	18.8	20.0
38. Physiognomic ear length	25.0	25.0	22.0	26.6
38a. Morphological ear length	19.5	20.3	21.0	29.4
39. Physiognomic ear breadth	16.0	18.2	18.7	16.0
39a. Morphological ear br. (otobasion superi-us-otobasion inf.)	20.5	20.0	17.6	21.5
40. Tail length	325.0[1] 28.0	for projection of hairs		370.0
41. Penis length	25.2	16.0	—	19.4
42. Subpubic angle to tip of penis	29.5	25.5	—	31.5
43. Diameter of penis	2.4	4.5	—	1.6
44. Subpubic angle to anus	13.2	13.7	18.3	5.0

[1] Taken on skinned body.

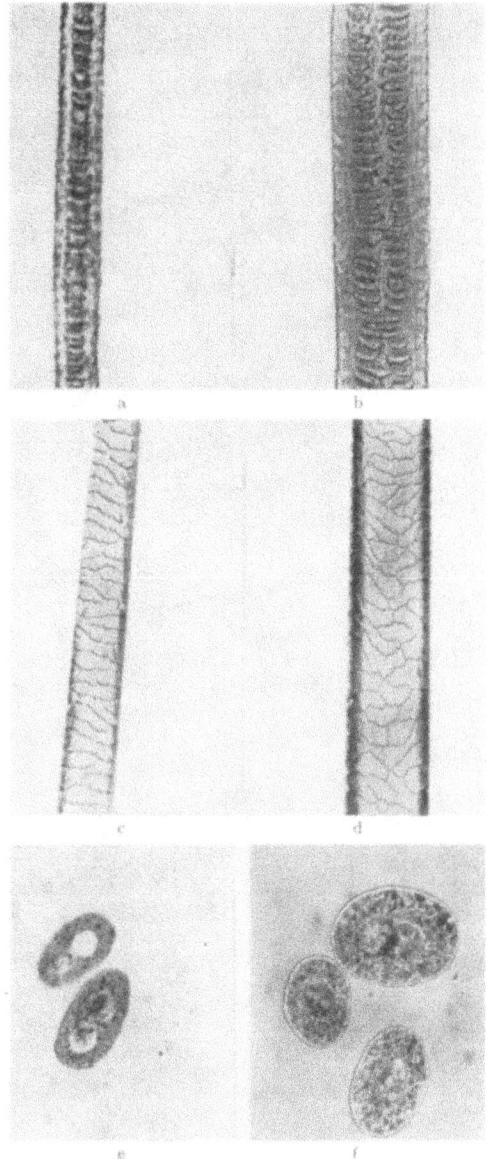

FIG. 2. Photomicrographs of hairs of *Callimico goeldii* and *Hapale jacchus* compared. *a.* Whole mount of hair from tail region of *Callimico* × 266. *b.* Whole mount of hair from tail region of *Hapale* × 299. *c.* Cuticular scale impressions of tail hair of *Callimico* × 266. *d.* Cuticular scale impressions of tail hair of *Hapale* × 266. *e.* Transverse sections (0.8μ) of two hairs from tail region of *Callimico* × 387. *f.* Transverse sections (0.4μ) of three hairs from tail region of *Hapale* × 387. (Courtesy of H. M. Appleyard, Wool Industries Research Association, Leeds, England.)

TABLE 3

SOMATIC INDICES

	PP77	T. tamarin
1. Rel. bisacromial diam.	30.6	35.7
2. Rel. bitrochanteric diam.	26	24.3
3. Rel. circumference chest	121	94
4. Thoracic index	84.5	80.5
5. Rel. bimammillary	—	80
6. Rel. pos. of nipple	—	83
7. Rel. pos. of umbilicus	38	—
8. Rel. l. upper limb	98.7	90
9. Humero-radial index	102	99
10. Forearm-hand index	86.4	83
11. Rel. l. pollex	18	21.25
12. Hand index	25.5	37
13. Rel. l. lower limb	110	94.8
14. Femoro-tibial index	106	102
15. Leg-foot index	93	94
16. Foot index	23	23
17. Intermembral index	95.5	95.7
18. Femoro-humeral index	71.7	68.5
19. Tibio-radial index	69.4	67.2
20. Foot-hand index	64	59.3
21. Rel. size of head	55.2	47.2
22. Head-trunk index	24.8	22
23. Cephalic index	60	71.7
24. Length-ht. index of head	42.7	43.4
25. Sagittal-vault index	74.5	72.8
26. Face-trunk index	15.4	15.3
27. Rel. size of upper face	16.9	10.2
28. Vertical cephalo-facial index	107.0	115
29. Upper facial index	45	31.1
30. Rel. nasal ht.	62.2	73.2
31. Rel. nasal br.	28.2	23.8
32. Nasal index	104	96.5
33. Rel. interocular br.	19.4	29.5
34. Ear index	64	60.2
35. Rel. size of ear	19.3	28

somatic indices derived therefrom according to the specifications and suggestions of Schultz (1921).

Principal results of the metrical comparison are (a) the general similarity in bodily size and form between *Callimico* and a typical tamarin, and (b) certain minor distinctions in relative proportions, notably the relatively shorter hand, the relatively shorter thumb, a longer pelvic limb in comparison with the spinal length, a relatively longer femur in comparison with the tibial segment of the hind-limb, a relatively larger head with more extreme dolichocephaly.

In the face the relatively smaller interocular distance and the relatively smaller ear of *Callimico* are to be noted. The intermembral proportions are strikingly similar in both animals.

EXTERNAL ANATOMY: REGIONAL DETAILS

1. RHINARIUM

Pocock's sketch accurately depicts the principal features. The external nose is extremely platyrrhine, with a septal breadth of 6.2 mm. The nares are rounded, having a diameter of 1.5 mm., and directed laterad. The internarial region is slightly depressed in the median line.

2. BUCCAL REGION

The buccal cleft when closed has the usual sinuous form met with in the neotropical Primates, i.e., with the median portion bowed upwards and the lateral parts descending. Lips are thin and deeply pigmented.

3. VIBRISSAE

A tuft of extremely short black bristly hairs occurs immediately above each naris (rhinal group). A few scattered curved hairs of similar nature but slightly longer spring from the upper lip. A few sparse hairs of doubtfully sinus-hair type are found on the lower lip, but there are no elongated or otherwise specialized supraorbital, genal, or interramal hairs.

4. PALPEBRAL REGION AND EYES

Eyelids are mobile and lax. Eyelashes are present on both lids but sparsely planted and extremely fine, particularly on the lower lid. Puncta lachrymalia lie 1.5 mm. away from the medial canthus.

A nictitating membrane is present but obscured by deep pigmentation. It is but 1.5 mm. across.

The cornea has a diameter of 6.55 mm., and is surrounded by a pigmented zone of sclera some 0.4 mm. in depth, beyond which the sclera is white, but the white cannot be brought into view during ordinary ocular movements.

The iris, tested by Ridgway's scheme, registers between hazel- and chestnut-brown. The pupillary opening in the fixed specimen has a diameter of 1.8 mm.

5. EXTERNAL EAR

The most striking feature is the broad line of attachment, which extends from the junction of anterior and superior portions of the helix inferiorly, in a vertical line to the extreme ventral limit of the aural margin. The distance from otobasion superius to otobasion inferius is therefore considerable (20.5 mm., i.e. 82 per cent of the physiognomic ear-height).

The next most important feature is the limitation of pigment to a broad marginal zone, slightly broader on the cranial than on the external surface of the organ. On the latter it includes most of the lamina distal to the antihelix, except near the otobasion inferius. It includes the posterior part of the helix, but not the portion near the otobasion superius (fig. 14).

In outline the pinna is roughly quadrangular, with rounded corners, but the superior margin rises slightly above the horizontal to gain the postero-superior angle. The posterior border is sensibly vertical, with but slight convexity; it passes by a broad curve into the almost horizontal inferior margin, but there is no flaring of the lamina postero-inferiorly such as occurs in *Tamarin* and *Marikina*, and no tethering of the inferior margin as is found in *Oedipomidas*.

The helix consists of two parts, an anterior broad vertical limb or crus, the anterior edge of which is

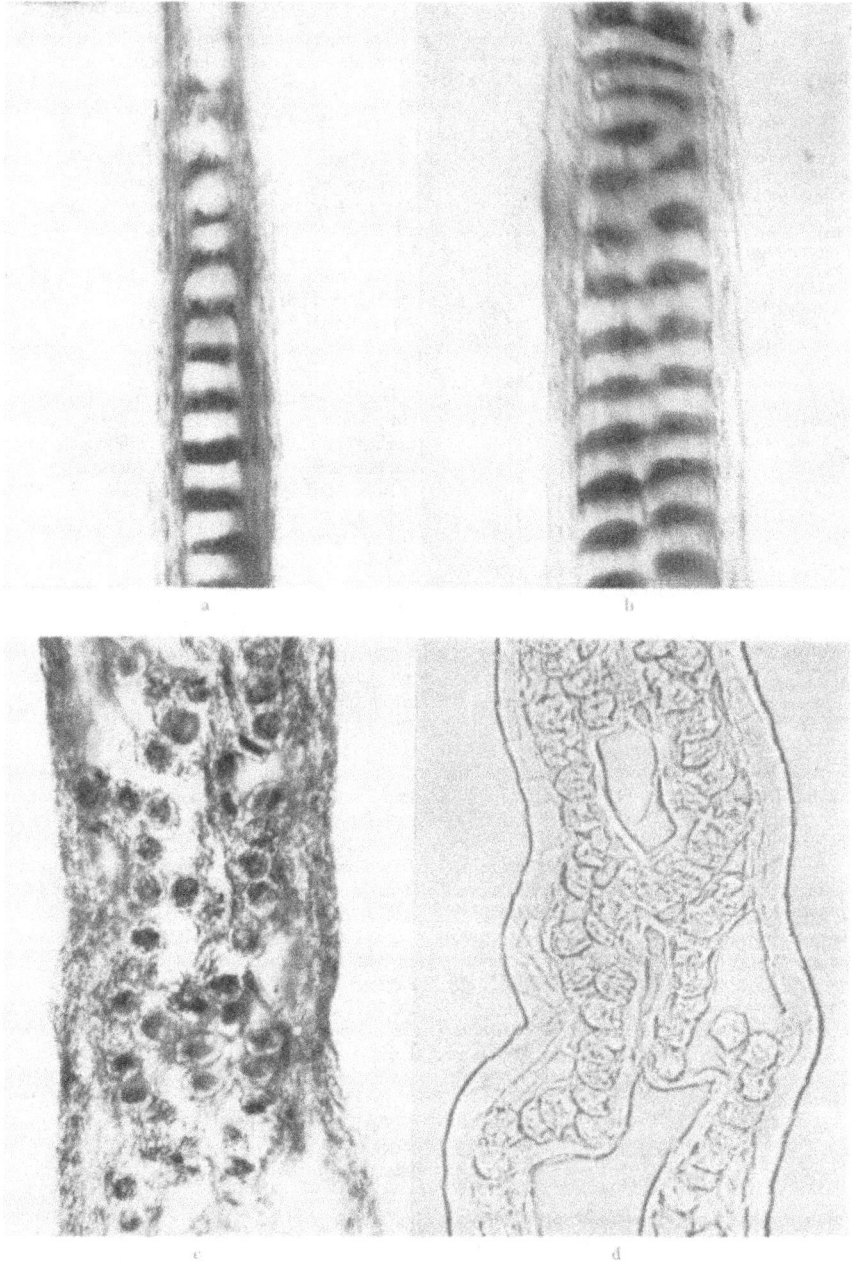

FIG. 3. Photomicrographs of hairs of *Callimico goeldii* and *Hapale jacchus* compared.　*a*. Longitudinal section of hair ($\times 7\mu$) from tail region of *Callimico* $\times$ 940.　*b*. Longitudinal section of hair ($\times 7\mu$) from tail region of *Hapale* $\times$ 940.　*c*. Hair of *Callimico* treated with KOH $\times$ 430.　*d*. Hair of *Hapale* treated with KOH $\times$ 430. (Courtesy of H. M. Appleyard, Wool Industries Research Association, Leeds, England.)

hidden beneath the skin of the cheek, and a superior portion which narrows, at first gradually, then abruptly to terminate before gaining the postero-superior angle.

The tragus is extensive in the vertical dimension but does not project much. It is separated from the antitragus by a deep intertragic notch. There is but a shallow depression between antitragus and antihelix. The latter curves evenly and ends above beneath the helix (at the junction of its two parts) without bifurcating. There is thus no fossa triangularis.

The pinna is practically naked, but a few black hairs spring from the lamina in a line parallel with the antihelix. They project a short distance beyond the free margin of the pinna. The area below the incisura intertragica also shows a few short white hairs.

6. CHEIRIDIA

Except for a few details, these are of typical hapalid conformation. Dollman (1937) states that the manus agrees with that in Hapalidae in the non-opposability of the pollex, but I find that this digit is set apart from the remaining digits to a greater degree than in any of the marmosets or tamarins. Moreover, the claws are robust, less acutely pointed, and less recurved than in any of the species of Hapalidae I have examined, which included members of all the eight recognized genera.

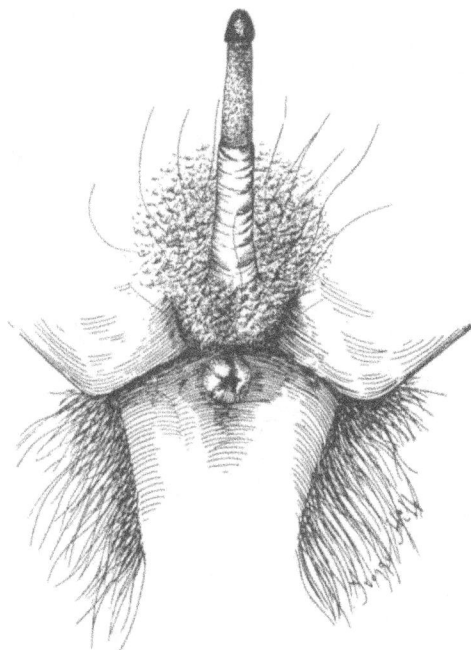

FIG. 5. Male external genitalia (from Hill, *Primates*, III).

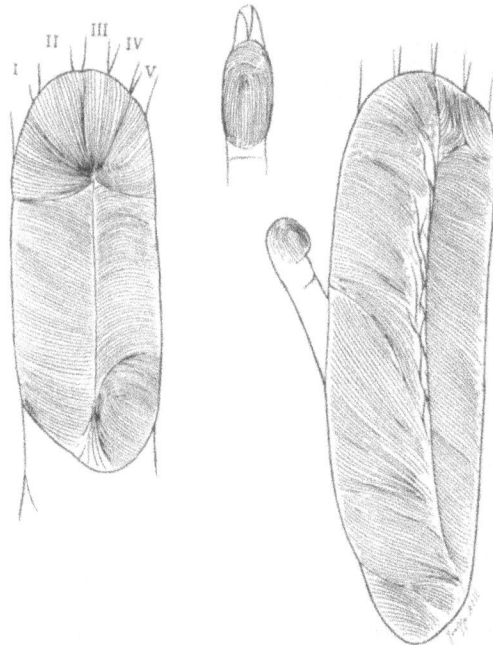

FIG. 4. Palmar and plantar dermatoglyphics of British Museum (Nat. Hist.) specimen No. 28.5.2.

The pigmentation of the manus has been dealt with above (p. 11). The digital formula of the manus is $III = IV > II > V > I$ and of the pes $IV > III > V > II > I$. The pollex extends distally as far as the first interphalangeal joint of II, but the hallux, though fully opposable, is extremely short, as in tamarins, not extending distally even so far as the metatarsophalangeal level. There is a moderate degree of interdigital webbing, which is more extensive between the pedal III and IV than elsewhere. There is nothing comparable with that which characterizes the genus *Leontocebus* ($= Leontideus$).

Palmar and plantar tactile pads are very feebly indicated, becoming flattened out and not recognizable even by their dermatoglyphic patterns, which merge with their neighbors, except in the center of the palm and sole, where indefinite markings are present. The papillary ridge patterns of the New York examples are shown in the accompanying figure, which indicates their simple character. There is evidently some slight individual variation, for diagrams prepared from the British Museum skin (no. 28/5/2) show slight differences in the pattern at the base of the hypothenar eminence of the manus and a simpler arrangement on the thenar pad. On the pes the pattern tends to be more transversely disposed on the heel region than in the one here depicted. No ridge patterns occur on the

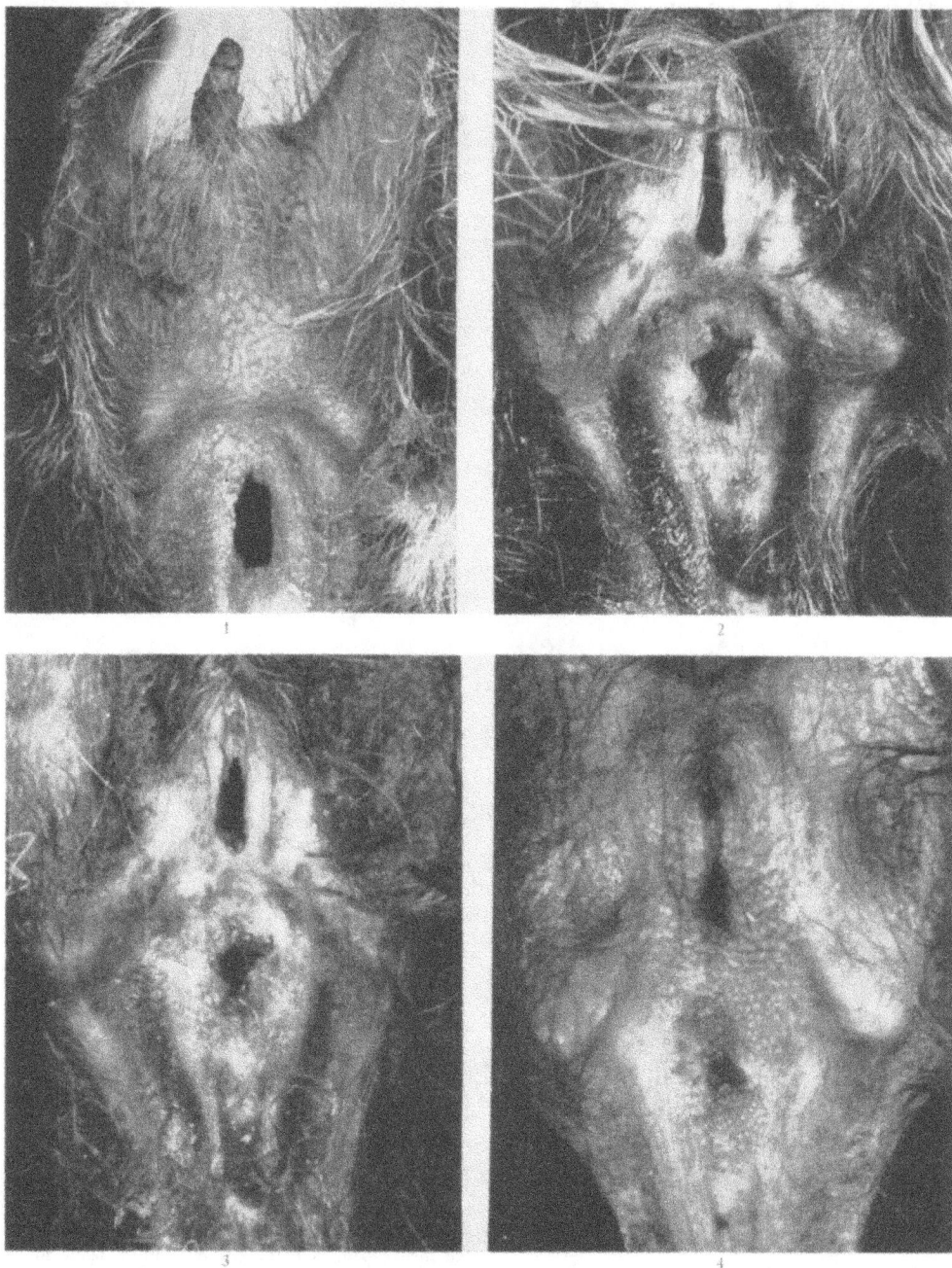

FIG. 6.   1. *Callimico goeldii* ♂ (Erikson's No. 975), external genitalia.   2. *Callimico goeldii* ♀ (Erikson's No. 974), external genitalia.
3. *Callimico goeldii* ♀ (Erikson's No. 974), external genitalia, labia everted to show clitoris.   4. *Callimico goeldii* ♀ (Erikson's No. 959), external genitalia, labia everted to show clitoris.
5. *Callimico goeldii*

proximal and intermediate phalanges of the digits, but on the apical pads it was noted that the central longitudinal ridges were fine in structure while the archlike peripheral ridges are coarse.

7. EXTERNAL GENITALIA OF MALE (FIGS. 5, 6, 51)

Pocock's account and figure, so far as it goes, are substantially correct. The penis is elongated and pendulous (for measurements see table 2). In the New York example the glans is deeply pigmented, exposed (presumably permanently in view of the pigmented character) and it would appear that there is insufficient laxity of the cutaneous covering to constitute a preputium, though Pocock's sketch indicates one well back from the glans. Pocock's animal—though stated to be adult—was probably immature, whereas the present animal, judged by skeletal and dental characters, is fully grown. A further distinction from Pocock's delineation lies in the greater length and more pendulous character of the organ.

In shape the glans is remarkably like its human counterpart, with a terminal vertical slitlike meatus and a well-defined corona glandis. But there is no frenulum praeputii, nor is a baculum palpable or visible radiographically. The sulcus retroglandis is very deeply incised, the corona overshadowing it, fitting caplike over the apex of the corpus penis. The corpus penis is cylindrical and pigmented, though less deeply than the glans. Its epithelium is marked out in more or less polygonal areas, but these bear no spicules or other adornments.

In the more mature male the retroglandic sulcus is extended to form a definite collum glandis, but this is visible only on traction, being normally hidden by the overlapping corona. The corpus in this animal is beset with numerous minute, proximally-pointing acutely pointed papillae arranged in approximately annular rows. These are, however, quite unkeratinized. Traces or vestiges of such papillae give a pock-marked appearance to the surface of the deeply pigmented glans.

Pocock's figure indicates a small globose scrotum similar to that of the tamarins of the genus *Tamarinus* and presumably containing the testes. This is curious in view of the suspected immaturity of the specimen, especially when, in the mature animal here studied, there is no apparent scrotal sac whatever. The scrotal site is occupied by a somewhat raised hairless and thickened area of integument extending from the attachment of the penis dorsad for some 10 mm. and differing in its browner hue from the duskily pigmented neighboring skin. It also differs in being marked by minute glandular nodules recalling those of *Tamarinus* rather than the grosser bodies seen in *Tamarin* or *Hapale*. This scrotal area is vaguely quadrangular in outline instead of subcircular; its appearance suggests its serving as a glandular (scent-producing) apparatus or friction pad comparable to that in the strepsirhine

genera *Perodicticus* and *Arctocebus* (Sanderson, 1940).

The testes in the New York animal are inguinal in position (fig. 50).

In the older male the testes are scrotal in position. Here the scrotum forms a distinct bilobed subpendulous sac depending some 21.8 mm. and a dorso-ventral diameter of 19 mm. The sac closely resembles a human scrotum in general shape but its surface is beset with numerous rounded glandular papules rendered more evident by virtue of a light pigmentary deposit between the almost contiguous papules. Papules are arranged in rows arching over the root of the penis, which emerges from the front wall of the sac near its neck. On the sides the papules form parallel rows coursing towards the fundus of the sac. Here they are larger than over the root of the penis. Papules also occur over the symphyseal area anterior to the scrotum, over a triangular area some 25 mm. sagittally, with the apex in front. At the base of the triangle the papules are minute and indistinct, but towards its apex they again enlarge. The pubic glandular nodules are not so clearly marked as the intervening reticular pigmentary pattern seen on the scrotum is here lacking.

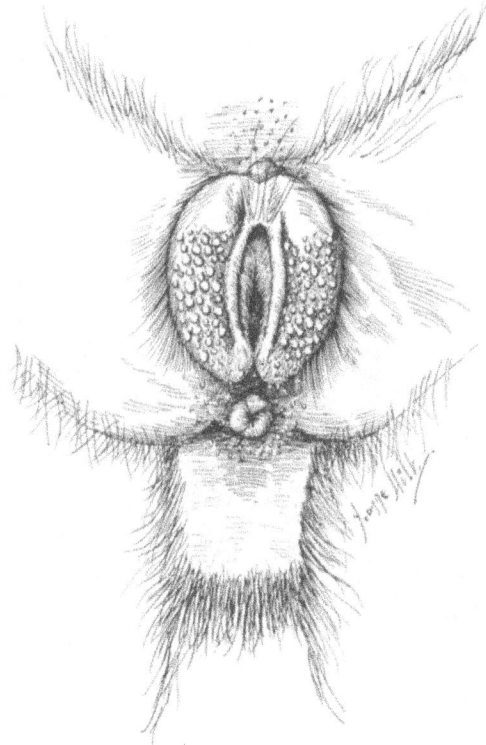

Fig. 7. *Callimico goeldii* ♀. External genitalia.

8. EXTERNAL GENITALIA OF FEMALE (FIG. 7)

These are shown in figure 6 which represents the organs in Erikson's two specimens. Figure 7 herewith is based on a mature female (no. PP169) examined fresh.

The vulva is of typically hapaline conformation. The rima pudendi, some 10 mm. long, is bounded by swollen labial masses marked everywhere, except for a narrow margin bordering the rima, by distinct rounded glandular papillae. These are arranged in vaguely longitudinal rows along the length of each labium and, as on the scrotum, are largest in the middle, becoming smaller towards the commissures. They are contiguous for the most part. Extensions of the glandular nodules occur both on the pubic region and the circumanal region. The pubic extension is triangular, resembling that in the male. The circumanal nodules are much smaller and more discrete. They cover the whole of the perineal body and extend around the anus on to the base of the tail, there being 4–5 rows dorsal to the anus.

The vulval mass measures 18 mm. long, 14.5 mm. broad and stands 8.0 mm. above the general level of the surrounding body surface.

The clitoris is inconspicuous, being represented by a bifid, smooth, pigmented glans at the extreme ventral end of the rima pudendi.

9. MAMMARY GLANDS

In the female PP169 the nipples are well developed, 5.2 mm. long and 2.5 mm. in diameter. They are unpigmented and spring from a zone of clear, almost hairless skin some 10 mm. below and 10 mm. behind the dorsal axillary fold. Mammary tissue is hypertrophied in this animal, forming a prominent margin ventrally some 10 mm. lateral to the mid-ventral line.

In the older male the nipples are similarly situated but smaller, but in the younger male they could not be detected.

THE SKELETON

A. AXIAL SKELETON

1. SKULL

A general description of the skull and comparison with that of other platyrrhines have been given in its several normae by Ribeiro (1940). The description

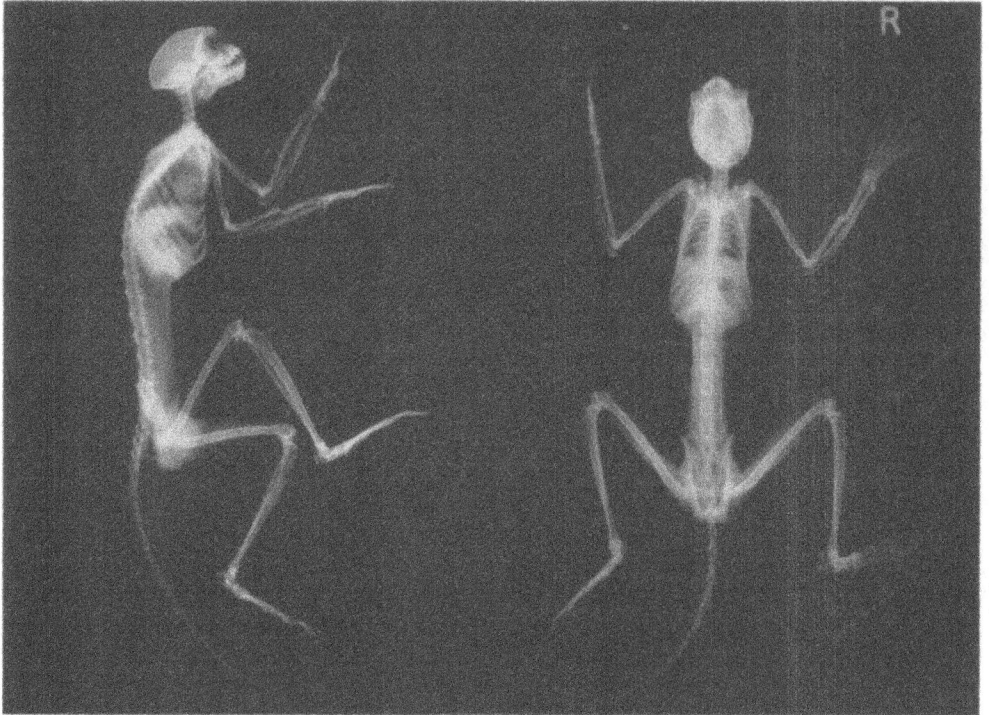

FIG. 8.   Radiographs of entire body of *Callimico goeldii* ♂.   (Courtesy of Mr. D. Stevenson Clark, Messrs. Ilford Ltd., London.)

given below is based upon the material in the British Museum, supplemented by observations on the freshly prepared cranium of the New York animal. It confirms in all essentials Ribeiro's findings and adds new data.

Though compared by Thomas to that of *Saimiri* and by Ribeiro to that of *Aotes*, as a whole the skull of *Callimico* is that of a tamarin, having the characteristic elongated, dolichocranial, parallel-sided brain case, with bulging occiput, large bullae, and the orthognathous face, with large circular orbital openings and small platyrrhine nasal opening. The mandible too is essentially hapaline: the only cebid feature is the dental formula (*vide* especially Ribeiro (*loc. cit.*)).

Elliot (3: 261) described the skull as "high" with a rounded brain-case, remarking on the feeble development of brow-ridges, the convex upper contour from nasals to occiput, the vertically expanded malar and the resemblance of the symphyseal region to that of the Cebidae. On the other hand he pointed out the similarity of the pterygoids to those of the marmosets but contrasted the more sloping plane of the orbital outlet.

Frontal Bone

From the glabella this extends back 23.5 mm. to the bregma, forming a more or less regular convexity in the sagittal plane, but somewhat flattened in the transverse. Rising some 30° for a few mm. immediately behind the supraorbital margin, the bone thereafter abruptly ascends for almost double that distance, forming a distinct forehead, then slopes more gradually to the bregma. There is no metopic suture, but its site is marked by a slight groove, with frontal eminences flanking it over the central areas of the bone, these being further defined by slight transverse depressions parallel to and some 2–3 mm. behind the supraorbital regions. The coronal suture is a fairly simple one; it describes an open V-shape, almost a U. There is some irregularity where the suture is crossed by the temporal line, the posterior edge of the frontal becoming suddenly more sagittally aligned for about 4 mm., and then reverting to its medial trend. This interruption is not, however, constant and is much less marked in B.M. No. 28/5/2/66.

On the internal surface there is a distinct groove for the sagittal sinus bounded laterally by raised bone. Lateral to this the bone is excavated, but the concavities do not correspond with the outward frontal eminences, being located more posteriorly near the coronal suture. There are also some dendritic depressions corresponding with the anterior tributaries of the sagittal venous sinus.

Laterally the frontal articulates with the parietal over its posterior three-quarters, the anterior quarter articulating with the malar in an almost horizontal line, slightly higher in front, where it cuts the orbital margin, than behind. Just above the middle of this fronto-malar suture a foramen leads into the orbit.

The fronto-nasal suture is an inverted V, sometimes asymmetrical, the right nasal rising slightly higher than

TABLE 4
CRANIOMETRY OF CALLIMICO

	mm.
Maximum length (muzzle—maximum occipital point)	51.5
Glabella—inion	45.4
Auricular height	17.9
Maximum cranial breadth (biparietal)	30.
Basion—bregma	21.7
Nasion—basion	32.3
Inion-opisthion	11.5
Inion-basion	16.3
Condylo-basal length (back of condyle to acanthion)	39.
Least frontal breadth	27.1
Biorbital breadth	30.5
Interorbital breadth	5.2
Bizygomatic breadth	35.5
Biauricular breadth	24.6
Bimastoid breadth	29.7
Bulla length	11.2
Bulla breadth	6.1
Orbital height	10.7
Orbital breadth	10.7
Nasal length (bone)	9.5
Nasal length (with opening)	14.0
Nasal breadth	5.1
Length of nasal opening	4.5
Palatal length	13.7
Palatal breadth (opposite M^2)	11.0
Breadth maxilla (opposite M^2)	19.8
Foramen magnum length	6.75
Foramen magnum breadth	7.0
Orbito-alveolar height	7.95
Olivary eminence—dorsum sellae	2.5
Across pituitary fossa	3.5
Optic foramen—plane of orbital outlet	4.7
Diameter optic foramen	1.7
Diameter olfactory foramen	1.8
Cranial capacity	10.5 cc.
Cranial capacity (of type female)	11.0 cc.
Upper tooth row, C–M^3	14.8
Upper tooth row, P^1–M^3	12.5
Mandibular length (with teeth)	34.7
Mandible—Coronoid height	21.6
Mandible—Condylar height	17.4
Bigonial breadth	22.7
Bicondylar breadth	30.
Depth at symphysis (oblique)	7.9
Height horizontal ramus (opposite M_1)	7.0
Minimum breadth vertical ramus	10.0
Lower tooth row, C–M_3	16.8
Lower tooth row, P_2–M_3	14.4

the left, and the frontal descending either side to form the uppermost part of the medial side of the orbital opening. Within the orbit the upper one-third of the medial wall is formed by frontal bone.

The orbital plates of the frontal are two thin but extensive laminae bulging into the cranium, highly concave on their orbital aspects and equally convex in the anterior cranial fossa. They meet in the median line except over a very small depressed area anteriorly, where the olfactory nerves proceed to the nasal fossa. The union extends back to the suture with the orbitosphenoid which is almost transverse, though a shade more anterior at its lateral end.

At the junction between the orbital plates and the main part of the frontal, the bone is excavated by extensive air-sinuses separated by a median septum which carries the olfactory nerves. Elsewhere there is no diploë or division of the bone into inner and outer tables.

Within the orbit, posterior to the fronto-nasal suture, the frontal articulates successively with lachrymal, ethnoid, palate, lesser wing of sphenoid and greater wing of sphenoid, the line of sutures being virtually horizontal.

Parietal Bones

These are extensive, taking the largest share in the formation of the cranial vault, for they extend back to the lambda, which is situated at the hindmost limit of the cranium; hence no part of the occipital bone is visible in norma verticalis.

In the median line of the vault the bone is raised to form a smooth crest, the sagittal suture passing first one side of this and then the other, but otherwise not highly denticulated. The crest, though sagittal in position, is not a sagittal crest in the ordinarily accepted sense, such as that normally occurring in *Leontocebus* (= *Leontideus*), i.e., it is not due to a fusion of the temporal lines of the two sides. Nevertheless, these lines are extremely well developed over the parietals, running parallel to each other some 4.5 mm. lateral to the median crest, though approaching each other a little more closely behind, before finally receding sharply downwards and laterad to the lambdoid crest, which they join some 7.5 mm. from the lambda.

On the cerebral aspect the sagittal suture is better marked. The groove for the sagittal sinus fades before reaching the anterior end of the parietals, but reappears on a larger scale on their posterior one-third. Here the sagittal suture lies along the bottom of the sulcus, which is bounded laterally by smooth raised eminences, flanked by concavities occupied by the parietal eminences of the cerebral hemispheres. Markings due to meningeal vessels are feeble.

Along its inferior border each parietal articulates with the squamosal by a ragged suture, the main direction of which is horizontal and located only a few millimeters above the auditory meatus. At the antero-inferior angle it meets the upper end of the small alisphenoid, which separates the squamosal from the extensive orbital plate of the malar. The latter articulates by a vertically disposed suture with the lower 8 mm. of the anterior border of the parietal constriction. An obtuse angle separates this part of the anterior border of the parietal from that involved in the coronal suture.

Occipital Bone

This is relatively flat since the nuchal surface and the opening of the foramen magnum occupy the same plane which makes a very obtuse angle (125°) with the plane of the basioccipital. Little of the bone is seen in norma occipitalis, which is largely formed by the parietals; most of the bone is visible only in norma basalis.

The nuchal plane is formed by the supraoccipital. It is bounded peripherally by the lambdoid crest, which corresponds with the line of the lambdoid suture. The crest terminates anteriorly on the mastoid region. Ventrally the supraoccipital is limited by the posterior edge of the foramen magnum. The external occipital protuberance corresponds with the central point of the lambdoid crest. The median part of the squama is marked by a smooth parallel-sided eminence, convex transversely but slightly concave from above downwards. This marks the site of the vermis cerebelli and corresponds with a concavity on the internal surface of the bone. The eminence is demarcated laterally, by sagittally directed grooves, from lateral eminences to which the deeper nuchal muscles (especially obliquus superior and rectus capitis dorsalis major) are attached. More peripherally are inserted the complexus and, superficial to this, the cleido-occipitalis, whilst on the median eminence, near the lambda is inserted the rectus capitis dorsalis minor.

The exoccipitals bear the occipital condyles. These are of oval outline, with their long axes directed obliquely forwards and medially. They lie on the anterior half of the lateral edge of the foramen magnum. Each is 6.8 mm. long and their anterior extremities are separated by a space of 3 mm. They are strongly convex longitudinally and less so transversely, and their dorsal extremities are slightly wider than their ventral ends. The pedicle of each condyle is perforated by the foramen for the XIIth nerve. Lateral to the condyle the exoccipital forms the jugular process, which is articulated with the temporal bone.

In front of the foramen magnum the basioccipital extends forwards between the tympanic bullae near the anterior ends of which a synostosis is formed with the basisphenoid. Immediately in front of the foramen the insertions of the ventral rectus muscles occur, and in front of these the pharyngeal roof is applied on smooth bone. The internal aspect of the basioccipital is smooth and concave both from side to side and antero-posteriorly. The large jugular foramen is an enlargement of the interval between the occipital and the temporal bones.

Temporal Bone

This consists, as usual, of squamosal, mastoid, petrous, and tympanic elements. The squamosal takes but a small share in the lateral cranial wall, being low, but antero-posteriorly extended, articulating in front with alisphenoid, above with parietal and behind with the occipital. The union with the alisphenoid descends to the level of the zygomatic arch, then turns posteriorly on the inferior aspect of the cranial wall, crossing the median part of the glenoid fossa to gain the lateral wall of the anterior end of the bulla postero-lateral to the hinder end of the foramen ovale.

The squamosal gives off from its lateral surface the zygomatic process. This is at first a dorso-ventrally flattened shelf provided with the usual three roots, anterior, posterior, and middle or descending. The first forms the posterior limit of the zygomatic fossa. The second is a posterior continuation of the zygoma on to the mastoid part of the bone, where it becomes confluent, at the asterion, with the temporal ridge. The intermediate or descending root forms the posterior boundary of the glenoid fossa and is raised superficially to form the post-glenoid process, which guards the antero-superior edge of the meatus. From the shelflike basis of the process the zygoma is continued forwards horizontally as a medio-laterally compressed bar, slightly arched, with an upward convexity. Its forward extremity overrides the upper edge of the zygomatic process of the malar, the suture uniting the two being elongated and virtually horizontal. The arch is bowed laterally and is visible in norma verticalis, i.e., the skull is phaenozygous as in *Hapale*, but to a lesser degree than in *Leontocebus* (= *Leontideus*).

The squamous and mastoid portions of the temporal bone meet above the external auditory meatus but no suture is retained here. The mastoid region is slightly inflated, but is not produced downwards to form a mastoid process. Its surface is roughened for attachment of fibres from the sterno-mastoid and trachelomastoid muscles. The suture with the parietal is extended back along the summit of the posterior root of the zygoma as it becomes continuous with the temporal ridge. At the asterion the suture turns ventrally at right angles to its former course, then runs forwards and medially to end at the jugular foramen. This part of the suture separates the mastoid from the supraoccipital and exoccipital. Internally the mastoid region is largely covered by the base of the petrosal, but the suture with the supraoccipital and exoccipital is traceable, though interrupted by excavations associated with the lateral venous sinus. The groove for this sinus is in places roofed in by liplike bony processes derived from the petrosal which, after bridging the groove, form sutures with the exoccipital on its opposite wall. The degree of bridging varies on the two sides in the skull of PP77, but on both sides the terminal part of the groove, as it approaches the jugular foramen, is extensively bridged. Possibly this is the arrangement referred to by Beattie in *Hapale* where he states that the foramen consists of two parts: "the anterior, which is often single, but may be subdivided into three foramina (an anterior, which transmits the inferior petrosal sinus, and a medial and lateral for the passage of the IXth, Xth, and XIth cranial nerves), and the posterior compartment for the transmission of the lateral sinus." His figure of the basis cranii interna, however, shows a lateral sinus unroofed throughout. No relic of the inferior petrosal sinus and no special opening for it at the jugular foramen were observed in *Callimico*.

The petrous temporal presents the usual three-sided pyramidal form. Its inferior surface is entirely hidden by developments of the tympanic, including the bulla,

to which the petrosal itself makes a contribution. Intracranially it presents a broad antero-lateral surface, narrowing anteriorly, where it is hollowed by the gasserian ganglion, and a narrower postero-medial surface marked by the internal auditory meatus, floccular fossa, and posterior semicircular canal. These two surfaces are separated by an extremely sharp border to which the tentorium is attached. The border is not uniformly straight or evenly sharp. Anteriorly it is slightly depressed for a few millimeters, the depression being limited in front by the posterior clinoid process of the basisphenoid and behind by a prominent, forwardly directed spicule of bone continuing the level of the main part of the crest. This depressed region gives passage to the stem of the Vth nerve. Posteriorly the crest rises to a summit over the site of the floccular fossa, then drops abruptly, becoming smoother before terminating in a backwardly directed spicule which overhangs the uppermost part of the groove for the lateral sinus.

The tympanic element is largely occupied in the formation of the external auditory ring—there is no osseous meatus—and the bulla. The bulla is a more ovoid swelling than in *Leontocebus* (= *Leontideus*), or *Hapale*, where it is described by Beattie as conical with a forwardly directed apex. An apex is present in *Callimico*, but it is a mere terminal outgrowth on the antero-lateral part of the main swelling and lies immediately behind the root of the medial pterygoid lamina, level with the apex of the petrosal. Just medial to this projection lies the opening of the tympanic tube together with that for the tensor tympani muscle. The long axis of the main sac is obliquely directed, passing from behind (level with the auditory opening) forwards and medially. Each bulla is 12 mm. long and 6 mm. in maximum width and, therefore, its attachment to the temporal bone circumscribes an oblong figure with rounded angles. A very slight transverse constriction divides it into a postero-lateral and an antero-medial moiety, the wall of the latter being further distinguished by the presence of air-cells (cellulae petrosae) whose limits can be determined through the semitransparent walls of the main chamber. This part of the bulla also differs in its reddish-purple color owing to the great vascularity of the mucous membrane lining the air-cells. It is derived from the petrosal element. The postero-lateral moiety forms the tympanic chamber proper and is slightly larger than the antero-medial portion. Its lateral wall is occupied by the auditory opening. A large foramen on the medial wall of the bulla, within the annular constriction, transmits the internal carotid artery to the interior of the bulla. The attachment of the lateral wall of the bulla to the cranial base is marked by a suture separating it from the zygomatic part of the temporal. Where the annular constriction meets this suture laterally is a foramen for the exit of the chorda tympani and a minute accompanying artery. The stylomastoid foramen perforates the temporal bone just behind the postero-lateral limit of the bulla.

The interior of the bulla was not studied as the skull was required by the American Museum of Natural History, but judging from the appearances seen through the outer wall and those ascertained by inspection through the auditory opening, there is no reason to suppose that there is any material difference from the account of the interior of the bulla of *Leontocebus rosalia* by van Kampen (1905).

Through the auditory opening can be seen, on the opposite wall of the chamber, the small opening reaching to the petrous air-cells. It lies medial to the passage leading from the tympanic tube. The promontory is also visible.

Sphenoid Bone

The basi- and pre-sphenoids are united to form a broad area of the basis cranii anterior to the two bullae. On the ventral surface the body of the sphenoid is smooth and flat and covered only by the roof of the pharynx. On the cranial aspect the bone is hollowed to form the hypophyseal fossa, bounded behind by the dorsum sellae, whose superior angles are projected laterally as posterior clinoid processes, beneath which the large oculomotor nerve is transmitted to the cavernous sinus. In front of the dorsum sellae the floor of the fossa slopes gently at the sides. Anteriorly the fossa is bounded by thickenings of the roots of the two orbito-sphenoids, separated by a median depressed tract running sagittally. From the anterior end of the presphenoid a vertical plate is continued forwards as in *Tarsius* (Woollard, 1925) and *Hapale* (Beattie, 1927b). This forms a septum between the two orbits in their hindmost parts and contributes a share to the medial wall of each orbit. It articulates inferiorly with the vomer and in front with the mesethmoid.

The alisphenoids or greater wings are large expanded bony plates consisting each of two parts set at right angles to each other, (a) an anterior, vertical, or orbital plate and (b) a posterior or horizontal plate. The orbital plate enters into the posterior wall of the orbit; it is a broad ellipse with pointed ends. The upper border unites with the posterior edge of the orbito-sphenoid, the two being set almost at right angles to each other, the orbito-sphenoid lying almost horizontally and the orbital plate of the alisphenoid virtually vertically. The horizontal portion forms the floor of the lateral part of the middle cranial fossa, articulating behind with the squamous temporal and laterally with the antero-inferior part of the parietal. It is perforated medially by the large foramen ovale; but there is no foramen lacerum medium between it and the temporal, and no foramen spinosum. At its apex the alisphenoid meets the frontal, but there is a foramen here leading to the upper and lateral part of the orbit. This transmits meningeal vessels derived from the ophthalmic, as pointed out by Mensa (1913) in *Hapale*.

The lateral pterygoid lamina is a quadrangular downward extension of the alisphenoid. Its root is formed by its upper border, which ceases at the anterior end of the foramen ovale. The anterior border articulates with the palate bone. The posterior border is very concave and meets the free slightly sloping lower border in a bluntly pointed subuliform projection, directed downwards and backwards. The lateral surface presents a distinctly forward and upward direction. On the medial surface is the minute medial pterygoid lamina or process projecting vertically downwards from the root of the lateral lamina and separated from it anteriorly in the rest of its extent by a notch. Below, the lesser plate ends in a blunt point. The mesopterygoid fossa between the two processes is very shallow as in Hapalidae.

The orbito-sphenoid (lesser wing) is fused medially with the presphenoid by two roots separated by the optic foramen. Laterally it forms a broad-based triangular plate which enters into the roof of the hinder part of the orbit. Its anterior border is articulated with the orbital plate of the frontal and its posterior border with the alisphenoid as described above. The latter union is deficient medially, where a triangular gap occurs constituting the inferior orbital fissure. It is separated from the optic foramen by the lower root of the orbito-sphenoid, which is a very slender bony spur.

Nasal Bones

Elongated and parallel-sided, but sometimes asymmetrical, these unite above with the frontals in an irregular transverse suture almost level with the superior orbital margin. They are united with each other by a median suture of wavy outline. Laterally they articulate with frontal (upper quarter) and frontal process of maxilla (lower three-quarters). The apex of the latter bone just gains the frontal, thus excluding the lachrymal from reaching the nasal. The lower border of the nasals is almost horizontal and receives the attachment of the nasal cartilages.

The nasal opening is broader than high. It is bounded above by the nasals, below by the premaxillae, and laterally, over a short extent, by the maxilla. The lower margin is U-shaped by virtue of the shape of the premaxillae (see below).

Premaxilla

On the facial aspect these form two cuboidal bony masses carrying the incisor teeth. They are united by a median suture, but above, in the lower part of the nasal opening, they are separated by a U-shaped notch, the bone, forming the alveolus of the upper central incisors, rising steeply on either side, afterwards sloping upwards and laterally more gently on the medial aspect of the anterior part of the corresponding nasal fossa. In the short diastema between the lateral incisor and canine the premaxillo-maxillary suture is seen, wavy in its lower course, where it is carried on to a thin crest. The crest continues as far as the lower limit of the corresponding nasal, but there is formed solely by maxilla, the premaxilla ending without reaching the nasal, though contributing somewhat to the lateral wall of the nasal fossa. The palatal portion of the bone is very limited chiefly on account of the large extent of the anterior palatine canals. The palatal part of the premaxillo-maxillary suture commences at the middle of the lateral border of the anterior palatine canal, running at first transversely laterad then bending forwards to escape through the diastema anterior to the canine.

Maxilla

From above downwards this is much compressed except at the inferior orbital margin, of which it forms the medial two-thirds. Here the orbital outlet narrows, hence the greater orbito-alveolar height compared with the distance between orbital floor and palate. The palatal aspect is extensive, accounting for that part of the hard palate posterior to the premaxillary contribution, with the exception of that derived from the palate bone, which occupies the posterior one-third medially, but not near the alveolar border, the whole of which from the canine onwards is formed by maxilla. Anteriorly the palatal process is notched by the posterior half of the anterior palatine canal.

On the facial aspect the maxilla gives off its ascending nasal process anteriorly. This articulates in front with the corresponding nasal bone. Above, it ends in a pointed apex which just gains the frontal, excluding the lachrymal from the nasal. The process forms the lower two-thirds of the medial margin of the orbital opening. Near its base it skirts the lachrymal foramen anteriorly.

Laterally almost the whole depth of the maxilla is involved in the broad area for articulation with the malar. A shallow notch is formed below between the malar process and the unaffected part of the "body" of the bone. The facial aspect of the maxilla is perforated just below the orbital margin and somewhat to the medial side of its center, by the single infra-orbital foramen. Anterior to this the bone is raised over the root of the upper canine. Posteriorly the maxilla ends in a rough tuberosity over the alveolus of M^3 above which it is separated by a deep transverse fissure from the orbital plate of the malar. Medially this part of the bone articulates with the palate bone.

The maxillary antrum is very small as in Hapalidae.

Malar Bone

A very large bone, chiefly from the extensive contribution made to the orbital wall, accounting for practically the whole of the lateral wall of that cavity. This orbital process is thin, highly convex externally and concave within the orbit. Above it articulates over a distance of 7 mm. with the lower border of the frontal, behind with parietal and alisphenoid, and below with the malar process and (more anteriorly) the body of the maxilla. Just wide of the articulation with the maxilla

the bone is perforated by two malar foramina and another foramen perforates the orbital plate near its posterior border. Lateral to the two former foramina the bone gives off its slender zygomatic process which extends backwards for 10 mm. It ends in a pointed tip which lies beneath the forwardly directed zygomatic process of the squamosal, the two being in contact, by means of an almost horizontal suture, over a distance of some 3–4 mm.

Palate Bone

This consists of the usual horizontal (palatal) and vertical plates, the latter being almost entirely within the orbit, as in *Hapale*, where it makes contact with the septal process of the presphenoid and forms the medial boundary of the inferior orbital fissure. Below this the bone makes contact with the pterygoids, serving to close the lower part of the mesopterygoid fossa. The horizontal plate forms the posterior one-third of the hard palate each side. The union with its fellow differs from that in *Hapale* and *Cebuella* in not forming any very distinct posterior palatal spine. To the posterior margin is attached a membranous sheet which forms the skeletal basis of the soft palate.

Lachrymal Bone

This is an oblong bony plate lying with its long axis vertically. It articulates anteriorly with the nasal process of the maxilla, behind with the os planum of the ethmoid, above with the frontal and below with the body of the maxilla. Its orbital surface is sharply demarcated into two areas by a vertical crest. Anterior to the crest the bone is hollowed, the depression deepening below and converted into a lachrymal canal by articulation with the maxilla. Behind the crest the bone slopes gradually to the general surface of the medial orbital wall.

Ethmoid Bone

This resembles closely that of the hapalids. There is neither cribriform plate nor crista galli, but a small depressed bony area at the anterior end of the middle cranial fossa which is perforated each side by a single olfactory foramen. Between these a median plate is continued downwards, forming the postero-superior part of the nasal septum, continued anteriorly as a cartilage. The lateral plates (ossa plana) are closely applied to the median plate, especially above, and there is no pneumatization of the bone. The bone separating the two orbits is quite translucent, for the nasal fossa does not extend upwards here to separate the median from the lateral plates of the bone.

Vomer

This agrees closely with that of *Hapale* as described by Beattie. It consists of a median plate which forms the posterior and inferior part of the nasal septum. Its free posterior border is obliquely disposed. Above the alae diverge to enclose the lower borders of the median plate of the ethmoid and the presphenoidal septum, but sufficient gap is left each side for the transmission of a small artery.

Mandible

The two halves, as in all adult Pithecoidea, are fused at the symphysis. They diverge uniformly therefrom at an angle of approximately 35°. In *Hapale* the angle is 45°, in *Leontocebus* 30°, but in *Cebuella* 50°.

The symphysis is sloped downwards and backwards from the incisor alveoli, more so on the lingual than the labial side. On the former it constitutes a simian shelf, below which the bone recedes slightly. Genial pits and tubercles are lacking.

The horizontal ramus is almost of uniform height throughout, but the lower border rises slightly towards the point of union with the vertical ramus, then descends again to the angular region, which is broad, truncate and directed backwards and slightly downwards, and very slightly inflected, but not quite as much as in *Leontocebus*.

The vertical ramus is low and broad, but the sloping anterior edge rises to a greater height at the coronoid process than the concave posterior edge gains in the condyle. The coronoid is retrorse and separated from the condyle by a deep sigmoid notch. The condyle scarcely shows a neck and is broadened transversely, projecting more on the medial than the lateral side.

From the anterior edge of the vertical ramus a strong bony crest delineates the anterior limit of the masseter on the outer face of the bone. The area for the medial pterygoid is also somewhat deepened by the inflection of the angular region.

The opening of the inferior dental canal lies beneath the middle of the coronoid process, not as in *Hapale* opposite the sigmoid notch. It is preceded by an oblique groove and is not shielded by a lingula, agreeing in this with *Hapale*, *Cebuella*, and *Leontocebus*, and differing from *Tamarin*.

The mental foramen on the lateral surface of the bone is single and located below the fore-part of the foremost premolar, about half-way across the height of the ramus.

Hyoid Bone

This consists of a median body of quadrate outline and two pairs of cornua, the whole constituting a U-shaped structure in the ventral median line between tongue and larynx.

The body is expanded in the sagittal diameter and compressed dorso-ventrally. It measures 5.5 mm. from side to side and 4.3 mm. cranio-caudally. The dorsal aspect is concave and the ventral convex and roughened for muscular attachments. The caudal border overlaps the cranial edge of the thyroid cartilage, thus hiding the thyro-hyoid membrane.

The greater cornu, 5.2 mm. long, is ossified separately and united to the body in such manner as to permit

limited mobility in a medio-lateral direction. It terminates in a rounded apex which forms a movable articulation with the summit of the superior cornu of the thyroid cartilage.

The lesser cornu is a shorter process lying craniad of the greater cornu and almost parallel with it. It is 3.0 mm. long, only the basal one-third, however, being ossified. The ossified portion is represented by a process fused with the corpus at its cranio-dorsal angle. The distal cartilaginous two-thirds is conical.

2. VERTEBRAL COLUMN

Vertebral formula: C.7, T.12, L.7, S.3, Co.3, Cau. 26.

This formula agrees, except for the greater number of post-sacral elements, with many specimens of *Hapale jacchus*, where, however, as Beattie (1927b) has shown individual variations occur at the thoraco-lumbar junction whereby on occasion 13 rib-bearing vertebrae occur. Some indication of instability is shown in *Callimico*, on the contrary, at the lumbo-sacral level, for there is some asymmetry here—considered in detail below. *Hapale* is given by Beattie as bearing 25 caudal vertebrae—using this to cover all the post-sacral elements. Hill (1957) records 28 for *Mico*, 30 for *Cebuella*, while Schultz and Straus (1945) give 33 for *Leontocebus* (= *Leontideus*) and 32 for *Oedipomidas*.

The spinal column displays well-marked curves, measuring, along the dorsal spines, 50 cm., 327 mm. of which are postsacral. Commencing with a ventral bowing in the cervical region, the thoracic portion follows with a distinct dorsal arch commencing with T.1 and reaching a summit at T.8. The arch is continued into the fore-part of the lumbar region, but beyond L.3 the spine tends to bow ventrad again, the ventral extent reaching its maximum with the last lumbar vertebra. The posterior two sacrals show a commencing upward trend which is continued through the coccygeal segments.

The curvature is more pronounced in respect of the vertebral bodies than with the spinous processes which, by virtue of their modifications in length in the different regions, help to smooth out the excessive curvatures. Thus in the cervical region and anterior thoracic region they are elongated, becoming shorter and stouter posteriorly in the thorax, and longer again opposite the lumbar curvature. T.9 is the anticlinal vertebra, as in *Hapale*.

As in *Hapale* the first postsacral vertebrae are freely movable, especially in a ventral direction, dorsal flexion being limited. Remaining postsacrals or true caudals are free to move in all directions.

Cervical Vertebrae: Atlas

This is the largest cervical vertebra, measuring 16.6 mm. transversely, and has the usual annular construction. The widest internal diameter of the ring is 6 mm., and the sagittal dimension is the same, so that the canal is approximately circular. This is brought about by the shelving inwards of the bone from the two lateral masses.

Both dorsal and ventral arches are compressed from above down and expanded cranio-caudally, but both are narrowed in the mid-sagittal line. In the case of the dorsal arch the narrowing is gradual and brought about by recession of both anterior and posterior borders. The ventral arch shows a distinct median notch posteriorly and a median ventral tubercle in front of this. In contrast, the dorsal arch shows no trace of a spine.

Lateral masses bear articular facets in front for the occipital condyles and behind for the axis. Condylar facets are elongated ovals with the long axes oblique and with the dorso-lateral third sharply angulated with the ventro-medial two-thirds. The lateral borders of the facets are convex and raised to form a sharp rim in front of the corresponding transverse process.

Facets for the axis are flattened, more rounded in contour, and directed backwards, medially, and slightly dorsad. Below they are separated by a narrow tract on the ventral arch from the smooth median area associated with the odontoid.

Transverse processes are short, compressed spurs of bone directed horizontally outwards from the lateral mass each side. Each presents two surfaces, antero-ventral (concave) and postero-dorsal (convex) and a rounded apex. The root is perforated by a vertebr-arterial foramen. A small foramen also perforates the dorsal arch near its anterior border immediately dorsal to the root of the transverse process.

Axis

This is a slenderer structure that that of *Hapale* as judged by Beattie's figure, having a longer, slenderer spinous process and less robust lateral masses. The body, when viewed from below, has a triangular outline, with the apex at the odontoid and lateral angles formed by the transverse processes. A sagittal ridge marks the mid-line of the centrum which is separated posteriorly from the transverse processes by a shallow notch each side. The odontoid shows the same sharp dorsal deflection as in *Hapale,* the angle of deflection being approximately 45°. The ventral surface of the odontoid is rounded. A groove separates the root of the odontoid from the atlantal articular facets. The latter are very slightly convex and of the same oval contour as their counterparts on the atlas. They face forwards, laterally, and slightly dorsad.

A slight forward spur marks the site of union of lamina and pedicle anteriorly on the neural arch. The arch slopes somewhat caudad and sends a flange-like articular process which rides over the forepart of the lamina of the third vertebra. The spinous process is laterally compressed and rises from the fore-part of the arch, being of lesser sagittal extent than the laminae.

It ends in a simple blunt apex, differing thus from that of *Hapale,* where the neural spine is bifid.

Transverse processes are conical, pointed, and directed obliquely backwards. Their roots are perforated by a vertebrarterial canal.

Cervical Vertebrae 3 to 7

Remaining cervical vertebrae are much alike. All have broad, shallow centra, the posterior borders of which overlap the anterior borders of the next in the series. Sagittal ridges mark the median line of all the bodies. In all the transverse processes are bifid, ending in dorsal and ventral tubercles separated on the cranial face of the process by a groove occupied by the corresponding spinal nerve and leading medially to the corresponding intervertebral foramen which is therefore bordered by a notch on the front of the pedicle of the neural arch. The ventral (costal) element on the transverse process of C.7 is reduced. All have flat, broad laminae which show an imbricated arrangement in the articulated spine. Neural spines are slender, not bifid, and increase in length from C.3 to C.7; they are directed slightly backwards, except that on C.7 which is almost vertical in position.

Thoracic Vertebrae

Bodies are more elongated sagittally than in the cervical vertebrae, but the first two are only slightly longer than cervicals. Thereafter the elongation increases with each vertebral body and is continued to an even greater degree through the lumbar series. Median ventral ridges are no longer present and the overriding at the posterior borders is not evident. At the cranial and caudal borders of each body is bilaterally a demi-facet for the head of the corresponding rib. The only exception is the last thoracic, which bears a whole facet for the head of the twelfth rib anteriorly and nothing on its posterior border. Pedicles are short, but wider in the sagittal plane than in cervical vertebrae, but the laminae, though oblong, are less so than in the cervical region, approaching a quadrate outline, especially in the hinder segments.

Spinous processes are elongated, especially anteriorly. They are laterally compressed throughout the series. Posteriorly they decrease gradually in height and they also change gradually in direction; on the first two thoracic vertebrae they are almost vertical; thereafter they incline more and more posteriorly until the ninth, which is anticlinal, after which a craniad trend develops and is continued throughout the lumbar series.

Transverse processes in the anterior part of the thoracic region are flattened dorso-ventrally and bear, on the inferior surface near their tips, facets for the tubercles of the corresponding ribs. Dorsal to these they present small tubercles for ligamentous attachment. In the hinder members of the series these tubercles become elongated cranio-caudally; this process commences in the eighth segment, increasing rapidly to the end of the series where the dorsal element presents an antero-superior and a postero-inferior tubercular thickening, the former homologous with the metapophyses and the latter with the anapophyses of the lumbar vertebrae. Facets for the costal attachment are lacking from the hinder two segments.

Lumbar Vertebrae

Elongation of the vertebral bodies in the sagittal axis is continued here, except on the last member of the series, which has a shorter, thicker centrum than either of the two preceding vertebral bodies. Each body is somewhat constricted in the middle, widening at each end. There is a distinct median ventral ridge, but this is faint on L.7. A small nutrient (venous) foramen lies to one side of the median ridge. Pedicles are short, but cranio-caudally elongated. Laminae are narrowed transversely but elongated cranio-caudally. Spinous processes are laterally compressed, cranio-caudally extended and directed forwards. Costal processes spring from the junction of body and pedicle; they are oblong, dorso-ventrally flattened blades with squared ends. They are very short on L.1, but increase greatly in length to L.6. They are directed obliquely forwards, downwards, and laterad at about the same angle with the spinal column as the corresponding neural spine.

Articular processes are borne by the laminae. Prezygapophyses bear concave articular facets on their medial faces, the postzygapophyses behaving in the reverse manner. Mammillary processes are developed as tubercles on the dorsal edge of each prezygapophysis. Accessory processes (anapophyses) are small tubercles developed near the dorsal side of the root of the costal processes.

The last lumbar vertebra is somewhat asymmetrical in regard to its costal elements, indicative of some morphological instability at the lumbo-sacral junction. On the right its costal element exceeds that of L.6 in length and breadth, but on the left it is short and attenuated, owing to inadequate ossification along its sacral border. The right half is therefore incipiently sacralized.

Sacrum

Three vertebrae are fused to form the dorso-ventrally flattened sacrum, which exhibits an oblong contour, being but slightly wider in front than behind. The articulation with the ilium is borne by the first two vertebrae and the contribution from S.2 is very slight. Two pairs of sacral foramina perforate the bone dorsally and ventrally at the sites of intervertebral intervals. The posterior pair are considerably larger than the anterior, both dorsally and ventrally. The bodies of the sacral vertebrae are convex transversely when viewed from below, the first two being flat in the sagittal dimension, the third slightly concave. Spinous

processes are fused to form a single, flattened blade of bone with a crenated dorsal margin. The lateral mass is expanded considerably in S.1 to form the major contribution to the sacro-iliac union. Posteriorly the costal element is reduced suddenly on S.2 to a dorso-ventrally compressed lamina which ends behind in a prominent angle directed laterally and slightly caudad. A small metapophysis is present on S.1 located medial to the first foramen close to the spinous process. Large prezygapophyses are present on S.1 with their articular surfaces facing medially. S.3 similarly exhibits a pair of stout postzygapophyses facing backwards and slightly laterad and downwards.

Coccygeal Vertebrae

The first three postsacral vertebrae much resemble the sacrals, but remain independent from each other. They are short and wide, thus differing markedly from the true caudals. They are provided with large, flattened transverse processes (smaller on the third) directed obliquely laterad and backwards in the same plane as the lateral mass of the sacrum; they expand at their lateral extremities. The neural arches bear spines and articular processes, but the former are very low, especially on the third member of the series. Ventrally these three vertebrae bear haemal arches at their posterior borders. These form independent chevron-bones connected to the corresponding vertebral body by syndesmosis. Short haemal spines cap the apex of the chevron and these are directed posteriorly on the first two vertebrae, while the third has a larger chevron bone with a forwardly directed haemal spine which leaves a gap of only 1.5 mm. between its tip and that of the backwardly directed spine of C.2.

Caudal Vertebrae

The first of this series is transitional, having a moderately elongated cylindrical body and short, triangular transverse processes at its rear end. It also has a neural arch and articular processes and a chevron-bone ventrally at its anterior end. The haemal spine on this chevron-bone is directed downwards and forwards like that on the preceding vertebra.

The next vertebra is greatly elongated and its general shape and proportions are repeated in the following nine vertebrae. Thereafter some shortening occurs, as well as diminution in diameter, these features being thenceforth progressive to the tip of the tail. A neural arch is present only on the vertebra mentioned at the beginning of this paragraph (i.e. Cau. 2). The ventral surfaces of Caudals 2 to 4 present median ridges. At their anterior borders are two paramedian tubercles— all that remains representing the haemal arch. There is a median dorsal ridge on Caudals 3 and 4 but this is lost on Caudal 5 and thereafter. Vestiges of transverse and articular processes are determinable as far as the tenth or eleventh caudal segment.

The terminal vertebra is a minute conical spicule 1.5 mm. long.

3. RIBS

Twelve pairs of ribs form the major part of the thoracic skeleton. Each terminates ventrally in cartilage. The anterior eight cartilages each reach the sternum directly, forming joints therewith. The following three pairs of costal cartilages narrow towards their ventral ends and terminate in a rounded apex which turns forwards, approaching the preceding cartilage and so gaining indirect attachment to the sternum. The two remaining ribs each side have small pointed cartilages which are embedded in the musculature of the flank.

The first rib is short, its ossified part being 11.0 mm. long, and cranio-caudally compressed, especially at its dorsal extremity. It articulates dorsally with body and transverse process of the first thoracic vertebra, ventrally with the manubrium sterni. Its cranial surface presents roughnesses for the insertion of the scalenus ventralis and medius, with a vascular groove between.

All the other ribs are medio-laterally compressed, but their outer surfaces are slightly convex in the horizontal dimension. Their angles are well marked and accentuated by roughnesses on the outer surface. Internally each rib presents, as in *Hapale*, a convexity near the cranial border and a concavity near the caudal border, giving the effect of each bone having a flanged hinder edge. Intercostal spaces average rather less than 3.0 mm. across.

4. STERNUM

This consists of a manubrium, body and xiphisternum. The manubrium is broad anteriorly, narrow behind, with a median ventral ridge and a forwardly directed presternal process. It presents a movable junction with the mesosternum. Its widest part is opposite the articulation of the first pair of ribs. Anterior to this each side is the clavicular facet.

The mesosternum consists of five segments, each somewhat dumbbell-shaped, the hindmost much shorter than the others but consisting of a single ossific element, i.e., not paired as reported by Parker (1868) in *Hapale*. It is followed by a longer, slenderer bony xiphisternum, which is capped caudally by an expanded xiphisternal cartilaginous plate (see also under Arthrology, p. 33).

B. APPENDICULAR SKELETON

1. CLAVICLE

Closely reminiscent of a human clavicle, the curves are more exaggerated and the bone lies more obliquely to the sternum than in Man. Maximum length is 24.3 mm. The junction of medial and lateral curvatures lies 14.5 mm. from the sternal end. The sternal articular facet is flattened and has an oval contour with the long axis placed cranio-caudally, 5.3 mm. long, in

contrast to the narrow (2.4 mm.) dorso-ventral dimension. In ventral view the sternal end narrows abruptly to the slenderest part of the shaft, which is at the commencement of the ventral bowing. The lateral part of the shaft is compressed cranio-caudally and broadened dorso-ventrally. The thoracic aspect of this part of the bone is somewhat concave for the reception of the subclavius muscle, and the stout coraco-clavicular ligament. The facet for the acromion is on the extreme tip, which faces laterad and slightly dorsad. It is covered with thick cartilage and shows no separate ossific center such as Parker (1868) recorded in one specimen of *Hapale*.

2. SCAPULA

Though triangular in general outline, this differs considerably from that of *Hapale* as described and figured by Beattie, chiefly owing to modifications along the cranial border and the suppression of the uncinate contour of the caudal angle. The bone is not adapted closely to the chest wall, lacking the usual ventral concavity; in fact the bone is remarkably flat.

The longest border is the axillary (35.2 mm.), then the vertebral (27.0 mm.), the cranial border being shortest (24 mm.). Cranial and vertebral borders meet in an obtuse angle (101.5°), vertebral and axillary in a very acute one (42°).

The cranial border is distinctly convex, especially in its ventral half. The suprascapular notch is converted into a foramen by ossification of the ligament, but this does not account for the convexity which is greatest dorsal to this structure. The vertebral border is rectilinear and the bone thicker here than in adjacent parts of the blade. The axillary border is also virtually rectilinear, but there is a slight heaping up of bone

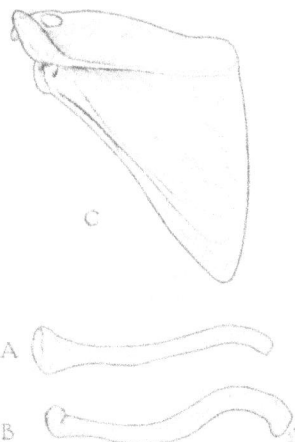

FIG. 9.　A. Left clavicle from cranial aspect.　B. Left clavicle from ventral aspect.　C. Left scapula from the dorsal aspect.

towards the caudal end of it. The bone is also thickened along this border, and in its ventral three-quarters it is grooved, so that the border is virtually double, the lateral lip being continued directly to the vertebral border, while the medial lip is solely responsible for the apical flange at the caudal angle.

The scapular spine stands out perpendicular to the main blade. It commences within 5.5 mm. of the angle between cranial and vertebral borders and traverses the blade approximately parallel to the cranial border. Its attachment reaches the neck of the scapula, reducing the great scapular notch to minimum proportions. Its free dorsal edge is only very slightly thickened. Towards the vertebral border, the height of the spine, which averages elsewhere some 3.5 mm., is abruptly reduced, sloping to nil at the border. A slight thickening of the edge is apparent at the point where this reduction takes place: this thickening is related to the insertion of the tendon of the trapezius. Laterally the spine expands into the acromion process, which is directed craniad.

The glenoid fossa is shallow with an oval outline, much narrower at its cranial than its caudal end. Its lateral border is evenly convex, but its medial border is distinctly concave (see also under shoulder-joint, p. 33). From the cranial pole of the glenoid fossa and adjacent part of the cranial border of the scapula springs the stout coracoid process. This consists of two parts, a short pedicle and a longer distal, flattened portion projected ventrad and somewhat laterally at right angles to the pedicle. The angle is joined by the ossified suprascapular ligament.

3. HUMERUS

A stout bone 54.5 mm. long. The head is a hemisphere separated by a neck from the two tuberosities. The latter are separated by the well-marked bicipital groove, whose lateral lip is the better developed and continued along the upper third of the shaft as part of the deltoid ridge. A strong median buttress is continued from the base of the lesser tuberosity, fading gradually about the junction of upper and middle thirds of the shaft. In profile the upper third of the shaft appears strongly retroflexed in relation to the remainder. This upper third is triangular in section, the middle third is circular and the lower third antero-posteriorly compressed. The radial nerve leaves no spiral groove on the bone. In the distal third the two epicondylar ridges become evident, increasing towards the epicondyles. A large oblique, smooth-lipped entepicondylar foramen is present—distinguishing the bone from that of *Hapale*, but aligning it with that of *Tarsius* and the lemurs.

On account of the torsion of the shaft, the transverse axis between the two epicondyles is rotated laterally so that the medial epicondyle is more anteriorly located than the lateral. The medial process is the more prominent. As in *Hapale*, it is grooved on the posterior side. On the articular surface, the trochlear and capi-

tellar areas are sharply demarcated, especially in front (see further under elbow-joint, p. 34). As in *Hapale* they lie in the same transverse plane (Shufeldt, 1914). Proximal to the trochlear surface there are a supra-trochlear fossa in front and an olecranon fossa behind, but the bone, though thin, is imperforate. A stout bar of bone separates the supra-trochlear fossa from the entepicondylar foramen; it is located proximal to the capitellum.

4. RADIUS

Contrary to the findings in *Hapale,* the radius is not appreciably more robust than the ulna. The chief distinction between them, apart from their extremities, lies in the cylindrical form of the shaft in the radius in contrast to its laterally compressed character in the ulna.

The bone is 53 mm. long and the shaft has a diameter of 2.6 mm. The disc-like head is 5.0 mm. in diameter and is subcircular in contour. The neck is long—the distance to the center of the bicipital tuberosity being 6.5 mm. From this level onwards the shaft is markedly bowed preaxially and also dorsally, and is slightly thickened at the summit of the arch, where pronator teres is inserted; the interosseous border is rounded. Distally the bone expands in the transverse dimension, becoming flattened on the volar side and convex dorsally where it is grooved by the extensor tendons. There is a stout styloid process on the preaxial side of the carpal articular surface.

5. ULNA

This measures 61 mm. long from olecranon to tip of styloid process of which 11 mm. is accounted for between base of olecranon and distal edge of great sigmoid notch. Half the latter distance is occupied by the extent of the notch.

The olecranon is a cuboidal mass of bone with large proximal surface for insertion of the triceps tendon. The great sigmoid notch is emarginate, especially at its proximal and distal borders. The lesser sigmoid notch is located on the radial side of the lower part of the greater notch and faces obliquely forwards as well as preaxiad.

The shaft is medio-laterally compressed, whence the dorsal surface is reduced almost to a border. The interosseous border is sharp. Both sides of the shaft are hollowed, especially proximally, where the excavations extend dorsal to the great sigmoid notch on to the olecranon. Distally the excavations become shallower and taper off to the rounded distal quarter of the bone. In *Hapale* only the preaxial surface is concave. The posterior surface is smooth, bowed axially, and convex from side to side. At the distal extremity the bone enlarges slightly, providing a convex surface for articulation with the lower end of the radius and a styloid process on the postaxial side.

6. CARPUS

This consists, as in *Hapale* and *Tarsius*, of nine bones. The proximal row comprises the scaphoid, lunate, cuneiform (triquetrum), and pisiform bones. A centrale (intermedium) is present. Distally occur the trapezium, trapezoid, capitate, and hamate. The pisiform is by far the largest carpal bone.

The *scaphoid* is of the usual form, with convex proximal articular area for the radius. It presents a tubercle on its volar surface; this is separated by a constricted neck from the body of the bone. Distally it articulates with trapezium and os centrale; postaxially with the lunate.

The *lunate* is also strongly convex proximally for articulation with the radius. It is concave on the preaxial side where it fits against the scaphoid. Postaxially it is convex in relation to the triquetrum. Distally it articulates with the head of the capitate.

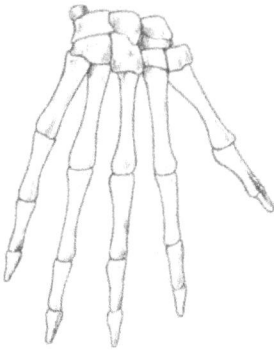

FIG. 10.   Bones of the right manus from the dorsal aspect.

The *triquetral* bone presents a quadrate surface towards the ulna, with which it articulates directly. Preaxially it presents a slight concavity for the lunate. On the ulnar side and somewhat ventrally it provides a facet for the pisiform. Distally it is apposed to the body of the hamate.

The large *pisiform* consists of two parts, (1) a basal portion articulating with the triquetrum and also directly with the ulna and (2) a distal rounded portion separated by a constricted neck from (1). This distal element probably ossifies separately (Sieglbaur, 1931; Ayer, 1940, 1948; Harris, 1944; Eckstein, 1944). It projects strongly in a volar direction thereby deepening considerably the carpal tunnel.

The summit of the pisiform is connected by stout ligaments to the hook of the hamate (pisi-hamate ligament) and the styloid process of the fifth metacarpal (pisi-metacarpal ligament).

The *os centrale* is a wedge-shaped element sandwiched between the scaphoid on the one hand and the trapezium (slightly), the trapezoid and the radial side

of the head of the capitate. As in *Hapale*, in contrast to the condition in *Tarsius*, the articulation with the lunate is very slight due to its displacement by the capitate. The bone is not larger than the trapezoid, as said by Beattie to be the case in *Hapale*.

Trapezium is projected on the volar side and barely visible from the purely dorsal aspect. It presents a characteristic oblique ridge. Distally it bears the usual saddle-shaped facet for the base of the first metacarpal.

Trapezoid is larger than the os centrale which articulates with the whole of its proximal surface. Dorsally it presents a quadrate outline. Radially it articulates with the trapezium and on the ulnar side with the body of the capitate.

The *capitate* is a large bone, elongated proximodistally, presenting a rounded head above and a cuboid body below. The head is proportionately small compared to that of the human os magnum; it articulates mainly with the lunate and os centrale. The body articulates with trapezoid, hamate and distally with metacarpals III and IV.

The *hamate* is a large bone with a subtriangular dorsal contour. Its ulnar border is projected ventrad as a hooklike process, connected by ligament with the pisiform. Proximally it articulates with lunate and triquetrum, preaxially with the body of the capitate; distally with metacarpals IV and V.

7. METACARPUS

The fourth is longest, third only slightly shorter, followed by V, II, and I in that order. All have relatively straight shafts, presenting the minimum of volar concavity in their long axes. The fifth presents a strong styloid process on its base. Their heads articulate with the carpals as in *Hapale* (*cf.* Hill, 1957: 134). There are intermetacarpal joints between the bases of II and III, III and IV, and IV and V. Distally their heads are held together by transverse intermetacarpal ligament.

8. OS INNOMINATUM

47.6 mm. long and greatest breadth from symphysis to ischial tuberosity, 20 mm. The elongation is largely due to the length of the ilium which measures, from crest to center of acetabulum, 31.5 mm.

The ilium is an oblong, medio-laterally compressed blade arranged somewhat obliquely to the sagittal plane, the gluteal aspect facing slightly ventrad. It presents three borders, anterior, dorsal, and ventral, its posterior end or "body" being continuous at the acetabular region with pubis and ischium. The anterior border forms the iliac crest; it is almost straight and vertically disposed. It forms rounded angles with the dorsal and ventral borders, the ventral (forming the anterior ventral spine) being the more marked. Dorsal and ventral borders are almost parallel. The dorsal is thick and slightly convex owing to the presence of the posterior dorsal spine or tubercle at about its mid-length. The ventral border is thinner and slightly concave. It presents, some 8 mm. anterior to the edge of the acetabulum, the posterior ventral spine, a large prominence associated with the tendon of the rectus femoris.

The medial surface of the bone presents the following areas: (a) a triangular dorsal area near the antero-dorsal angle which is occupied by muscular attachment (erector spinae), (b) an area posterior to this receiving the ilio-lumbar ligament which connects the bone with the tip of the costal process of the last lumbar vertebra, (c) the auricular surface for articulation with the sacrum; this occupies almost the whole depth of the blade, but receding dorsad in its hinder portion, (d) the iliac "fossa" which, far from being hollowed, is in fact convex in the dorso-ventral direction. The surface is crossed obliquely by the ilio-pectineal line, which extends from the postero-ventral edge of the auricular surface backwards and ventrally on to the forepart of the pubis, where it is more sharply indicated.

The lateral or gluteal surface is strongly concave both sagittally and dorso-ventrally.

The acetabulum is circular in contour, deepened by a fibro-cartilaginous labrum. This closes the notch which is on the posterior margin. The notch is continued as a groove for some distance on the surface of the body of the ischium. The acetabular floor is very thin.

The ischium is a stout bar of bone continued backwards beyond the posterior wall of the acetabulum. It is very thick dorsally, thinning considerably towards the obturator foramen, of which it forms the upper boundary. Near its root it bears dorsally a broad, low but blunt, ischial spine. The bone ends behind in the somewhat upturned ischial tuber. From its hinder half the bar gives off ventrally the ischial ramus, a thin blade of triangular outline with its apex below meeting the pubic ramus. The posterior border of this is vertically disposed and thicker than the remainder.

The pubis presents an anterior ramus, descending obliquely from the inferior wall of the acetabulum and a posterior ramus proceeding backwards around the obturator foramen to meet the ischial ramus as already indicated.

The posterior ramus meets with its fellow in a long symphysis. A sharp pubic angle marks the site of union of anterior and posterior rami.

The obturator foramen is almost circular, with a diameter of 9.0 mm.

9. FEMUR

This is a relatively straight bone, 75 mm. long in its maximum trochanteric length, and only 1.0 mm. less in its oblique length from caput to condyles. The hemispherical head rests upon a short, antero-posteriorly compressed neck. Three trochanters are present. The greater is a somewhat cuboidal bony mass with a broad, rough lateral surface which narrows below into an oblique ridge which proceeds downwards and backwards

on the shaft, presenting the small third trochanter at its lower end. Medially, on the posterior surface, the great trochanter is separated from the neck by a deep vertical groove representing the digital fossa. The lesser trochanter is very stout, with an axially elongated base and a broad summit directed medially and slightly posteriorly. Intertrochanteric lines are not clearly represented, nor is the linea aspera. The shaft is, however, subtriangular in section owing to some posterior buttressing, but the surface is everywhere smooth. There is a minimum of antero-posterior bowing. Below, the shaft widens gradually, in the transverse dimension, but the popliteal surface is not particularly flattened, being indeed slightly convex transversely. Supra-condylar ridges are not marked and there is no well-defined adductor tubercle such as Beattie describes in the marmoset.

Of the two condyles, the lateral is the more bulky, but the medial is longer antero-posteriorly. The patellar groove is shallow and the intercondylar notch 1.7 mm. across. Torsion of the shaft as judged by the difference between the transverse axes of the upper and lower extremities of the bone amounts to about 30° as in *Hapale*.

10. PATELLA

This is a small bone about 5.0 mm. in diameter, subquadrate in outline with rounded angles. Its anterior surface is convex in both directions, and its posterior cartilage-covered surface convex from side to side and concave from above downwards. Proximal to it is a cartilaginous suprapatella.

11. TIBIA

The maximum condylo-malleolar length is 75.5 mm.; from condyles to inferior articular surface the distance is 71.2 mm. It is thus, as in *Hapale*, slightly longer than the femur. The proximal articular surface much resembles the same area in Man; it is considered below under knee-joint (p. 35).

The shaft is medio-laterally compressed and sagittally bowed, especially in its upper half. The posterior surface is reduced to a broad, rounded border and the bone, in section, loses therefore much of its usual triangular form. All ridges and borders are feebly marked, except for the cnemial crest descending 12 mm. distally from the anterior tubercle. In the lower third the bone is approximately circular in cross-section.

The lower extremity broadens transversely into a pyramidal form with anterior, posterior, and medial surfaces, the last-mentioned being prolonged distad as the medial malleolus. At the junction of medial and posterior surfaces an elevated ridge is formed with a groove immediately laterad for the flexor tendons. On the anterior aspect the malleolus is separated by a notch from the adjacent part of the distal border. The facet for the inferior end of the fibula is slightly concave antero-posteriorly.

12. FIBULA

This is 71 mm. long, straight and slender. It takes a somewhat oblique course with reference to the tibia, its proximal end being located more posteriorly than the distal. Above, its head articulates with the inferior aspect of the lateral tibial condyle. The shaft is cylindrical, showing scarcely any borders; except the interosseous border which is only feebly marked. The distal end broadens transversely, presenting a flat posterior surface and a convex anterior one. It terminates as the lateral malleolus, which descends as far as the medial malleolus.

13. TARSUS

Tarsal bones are the same in number and general arrangement as in Man.

Talus

This comprises a body and rounded head connected by a constricted neck. It has an overall length of 10.5 mm. and maximum antero-posterior extent of 10 mm. The breadth of the body is 6.9 mm., height 5 mm., and diameter of head 4.5 mm. The angle subtended by the sagittal axis of the body with the axis of the neck and head is 24°.

The body is almost entirely articular, presenting (1) a convex trochlear surface above adapted to the inferior articular facet on the tibia, (2) small, comma-shaped medial facet for the medial malleolus, (3) larger comma-shaped facet laterally for the lateral malleolus, and (4) an oblong facet with rounded angles and concave for the articulation with the posterior facet on the calcaneus. The deepest part of the depression forming this last-mentioned concavity forms a groove directed in the same axis as that of the head and neck.

The neck is long (3.0 mm.) and constricted. Above it presents a rough area for ligamentous attachment (see under Joints). Below it is smooth, being covered by articular cartilage extending proximally from the head. The head is not spheroidal, but broadened transversely, its longest axis being slightly oblique, the lateral pole being somewhat more dorsally placed than the medial. Below, as already noted, the articular cartilage encroaches upon the neck, leaving a small area behind, between the articular area and the posterior calcaneal facet, a non-articular area which forms the roof of the sinus tarsi and which is occupied by the interosseous ligament.

Calcaneus

A greatly elongated bone, 14.2 mm. in total length of which 4 mm. projects behind the posterior limit of the talar facet and 6.8 mm. posterior to the hind edge of the large medially projecting shelf (sustentaculum tali). Maximum breadth opposite the sustentaculum is 5.8 mm.; maximum breadth, omitting the sustentaculum, 4.1 mm.; maximum height 6.1 mm.

The posterior extremity is oblong, with the narrow diameter transversely. The major part of it is related to a bursa beneath the tendo Achillis and is covered with smooth cartilage; this is concave in the transverse dimension and consequently the medial and lateral borders are raised, sharply demarcating the posterior surface from the sides of the bone. The tendon is inserted only into the inferior quarter of the bone, which is here roughened.

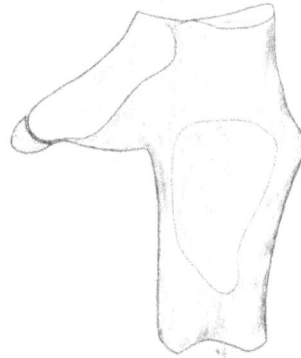

FIG. 11.   Left calcaneus from the dorsal aspect × 4.2.

The upper surface presents two articular areas, both for the under surface of the talus. The posterior is triangular in form with the apex posteriorly. It is convex sagittally, but fairly flat in the coronal plane. Behind this facet the bone is non-articular. In front of it is the sulcus tarsi occupied by the thick interosseous ligament. Medial to this area is the sustentaculum tali, bearing on its upper surface the concave anterior facet for the head of the talus. This facet has a dumbbell-shaped contour and its long axis is directed from behind forwards and laterally. Medial to its posterior moiety is a sesamoid cartilage within the spring ligament. The under surface of the sustentaculum is grooved by the tendons of flexores digitorum tibialis and fibularis.

A strongly marked peroneal tubercle occurs on the lateral surface of the calcaneus at the level of the hinder end of the sustentaculum. As in *Hapale,* this is smooth-surfaced and presents no groove below.

The anterior end of the bone is wholly occupied by an articular facet for the cuboid bone. Rounded in contour with dorso-ventral and medio-lateral diameters subequal, it is slightly concave and notched below from encroachment of the inferior calcaneo-cuboid ligament.

Os naviculare

This presents the usual proximo-distally compressed form with articular cartilage on its anterior and posterior surfaces. The posterior facet is concave in adaptation to the head of the talus and is of evenly oval outline—not pyriform as reported by Beattie for *Hapale.*

The long axis of the oval passes from above downwards and medially, but the minor axis is but little shorter than the major. There is no articulation with the calcaneus, but laterally there is a union with the cuboid, the facet being confined to the upper part of the navicular and of quadrate outline. The lower part of the surface is rough for ligamentous attachment. The upper and medial surfaces pass imperceptibly into each other, forming a sweeping convexity from above downwards and medially. In the sagittal plane the lateral surface is slightly concave, owing to the raised proximal and distal borders. The tuberosity is present, blunt, and rounded.

Distally the navicular articulates with the three cuneiforms. Collectively, the three facets describe an inverted L-shaped outline, with the longer limb lying vertically at the medial side of the foot and formed by the combined facets for ento- and meso-cuneiforms. The entocuneiform facet is slightly convex and the mesocuneiform area flat. The two are set at a very slight angle with each other. The facet for the ectocuneiform forms the short limb of the L and is rounded, especially below, where it forms a sharp inflexion with the lateral edge of the long limb of the L. Dorsally the angle between short and long limb is contrastedly rounded off. The surface of the ectocuneiform facet is slightly convex.

Cuboid

The proximal aspect of this bone is wholly concerned in articulation with the calcaneus, and therefore covered with articular cartilage throughout. Its outline is reniform, with the notch inferiorly. Medially the bone presents facets for the navicular and ectocuneiform bones. Dorsal, lateral, and inferior surfaces are rough for ligamentous attachment. Distally there is an undivided flat, articular cartilage-covered surface for the fourth and fifth metatarsals.

Cuneiform Bones

Entocuneiform is the largest and mesocuneiform the smallest of the three. Proximally they are entirely cartilage-covered, the facet on the entocuneiform being concave, the others flat. All are broad dorsally and narrow ventrally, the two lateral bones being particularly wedgelike. Distally they present an irregular line of articulation with the metatarsus, the mesocuneiform being shorter than the others, with consequent revêtement of the second metatarsal base.

Distally the entocuneiform presents a saddle-like articular area for the hallucial metatarsal. In general outline this facet is oval, but there is a notch on its dorsomedial border and a corresponding bulge on the opposite border. The major axis of the oval is oblique, passing from above downwards and medially at an angle of approximately 45° to the transverse axis of the metatarsus. The two poles present distally each a rounded convexity, separated by a shallow sulcus perpendicular to the major

axis of the oval. The groove joins the dorso-medial notch. Towards the plantar side of the facet, the groove disappears, the ventro-lateral bulge being convex.

14. METATARSUS

With the exception of the first, all the metatarsals are long, slender bones. The three postaxial ones are much alike in size and the distal ends (heads) of all three project to approximately the same level. The second is peculiar in its being almost 5.0 mm. shorter and in its strikingly much more slender shaft. The head of the second metatarsal barely surpasses the distal end of the

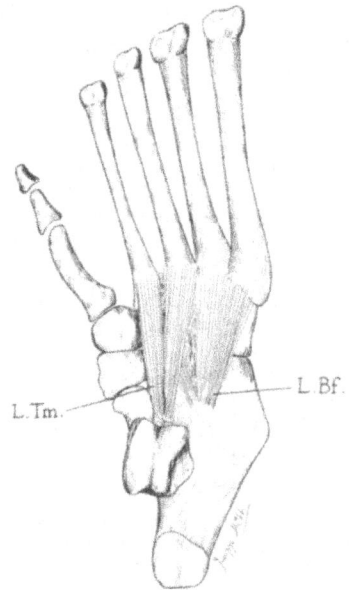

FIG. 12.   Bones and ligaments of the left pes
from the plantar aspect.

hallux when the latter is placed alongside. The first metatarsal is short and stout and presents its own peculiarities at its base, where its articular surface conforms to that on the entocuneiform. The fifth is peculiar in its enormous styloid process, which projects proximally on the lateral side of the foot almost as far as the calcaneo-cuboid joint. This arrangement considerably restricts abduction of the fifth toe.

Sesamoid bones occur in pairs in the flexor tendons opposite *all* the metatarso-phalangeal joints.

15. PHALANGES

The only feature for mention here is the shape of the terminal phalanges. These resemble those of *Hapale* and other genera of Hapalidae on all toes being acutely pointed except on the hallux, which bears a dorso-

ventrally flattened, spatulate terminal phalanx in correlation with the flat nail borne thereon.

ARTHROLOGY

1. TEMPORO-MANDIBULAR JOINT

A meniscus divides the joint into upper and lower
cavities. It is 4.8 mm. broad and 2.8 mm. sagittally.
Its upper surface is fairly flat, but the lower strongly
concave antero-posteriorly. Anteriorly a broad attachment (3.4 mm.) of fibres from the lateral pterygoid
muscle takes place; elsewhere the meniscus is firmly
attached to the capsular ligament. The latter is very
thick on the lateral side; thinner posteriorly and very
thin medially. No local ligamentous thickenings were
discerned.

2. SUBOCCIPITAL JOINTS

The membrana tectoria is a broad membranous sheet
covering the dorsal aspect of the suboccipital joint system and differs in no important respect from the description given by Hecker (1922) of its appearance in *Cercocebus* and *Cercopithecus*. It is attached cranially to the
ventral margin of the foramen magnum and caudally to
the dorsal surface of the body of the axis vertebra. Beneath it lies a cruciform arrangement of ligaments composed of the median occipito-odontoid ligament, the two
lateral oblique check-ligaments (lateral occipito-odontoid
ligaments) and the transverse ligament passing across
the neck of the odontoid. There is no evidence of the
superficial cruciate stratum mentioned by Hecker in
the higher catarrhines. The oblique ligaments are very
stout, arising below from the sides of the head of the
odontoid and inserting at the sides of the ventral border
of the foramen magnum near the medial end of the
articular facet on the condyle. The fibres are contiguous with those forming the capsule of the atlanto-
occipital articulation at this point. The transverse ligament is feebler than the oblique ligaments, but has the
usual attachments. It shows no tendency to ossify
(*cf. Tarsius;* Hill, 1955).

The antero-ventral parts of the capsule of the two
atlanto-occipital joints are connected by ligamentous
tissue contiguous ventrally with the ventral common
ligament of the vertebral bodies (ventral atlanto-occipital membrane). A synovial-lined recess is formed
laterally between the edge of the condylar facet of the
atlas and the capsular attachment.

3. THORACIC JOINTS

Costo-vertebral articulations call for no special remark, nor do the costo-chondral junctions. Chondro-
sternal unions are relatively mobile, with incipient
synovial cavities. The ventral ends of the cartilages
are convex and accommodated within concavities formed
at the sites of junction between the sternal segments.
Radiating ligamentous bands connect the skeletal elements both dorsally and ventrally.

The manubrio-gladiolar joint is a well-developed
synovial joint of the ball and socket variety, the convexity being formed by the manubrium and the socket
by the first stenebra of the mesosternum. Movable
joints also occur between the several pieces of the
mesosternum and between the last piece and the bony
part of the xiphisternum. The most anterior is incipiently synovial, but the articular surfaces are flat.
The remainder are lax synchondroses with incipient
cavity-formation in the center.

4. STERNO-CLAVICULAR JOINT

This is a well-developed synovial joint with two
completely distinct cavities separated by a meniscus.
The capsule is thin and lax, especially dorsally, where
the head of the clavicle forms a rounded prominence.
Ventrally the capsule is supported by fibres derived
from the tendon of the sterno-mastoid and by some from
the pectoralis major. The sternal facet is relatively
flat, but the clavicular element is subdivided into two
convex parts by an oblique groove running mainly
dorso-ventrally. The caudal prominence is the more
convex and takes a larger share in the activity at the
joint. The meniscus is thicker dorsally than ventrally
and there is at the dorsal end of the groove, in the angle
between the two subdivisions of the articular area, a
thicker localized ligament—possibly the remains of an
interclavicular ligament, but of which there is no other
trace as there are no fibres traceable across the middle
line.

Movements are very free in the dorso-ventral plane,
but limited in the sagittal plane.

5. ACROMIO-CLAVICULAR JOINT

In contrast to the preceding, this is somewhat degenerate. It has lost its synovial character and the
meniscus has been reduced to an interarticular fibro-
cartilaginous pad of some thickness, but attached on
both sides to the adjacent bone. Movement is, however, fairly free, especially in the dorsoventral direction; but it is more limited in the sagittal plane.

6. SHOULDER JOINT

The circumference of the glenoid cavity is an
elongated oval with the cranial end much more acute
than the axillary. It measures 7 mm. in the long axis
and 4.5 mm. across at its widest part. The ventral
border is very distinctly notched at the junction of the
coracoid and scapular contributions. The humeral
head exhibits a considerably larger articular surface
than the glenoid, hence movement is free in all directions.. Its diameter is 7.2 mm. A small zone (1.2 mm.
across) of the neck of the humerus is intra-articular on
the inferior aspect of the head; this is covered with
synovial membrane.

The capsule of the shoulder joint is everywhere thin;
there are no local gleno-humeral ligaments. Support is
afforded by the tendons of the short scapular muscles

and by the extensive coraco-humeral ligament which has a broad attachment along the whole lateral border of the coracoid. There is no special region where the capsule is more lax than elsewhere. It is attached proximally to the labrum glenoidale and distally to the morphological neck of the humerus immediately beyond the articular cartilage margin—except below as indicated. Between the tuberosities the capsule is attached to the transverse humeral ligament or retinaculum which bridges the gap and holds the biceps tendon in position.

A deficiency in the capsule, permitting an escape of a synovial bursa beneath the subscapularis tendon, occurs opposite the notch on the ventral margin of the glenoid, near the scapular attachment of the capsule.

The long tendon of the biceps is intra-articular, arising from the most cranial point of the glenoid margin; it is covered by synovial membrane, which sheath is prolonged into the bicipital groove.

7. ELBOW JOINT

A hinge joint incorporating articular areas on the lower end of the humerus and proximal ends of radius and ulna. On the humerus the trochlear surface is quadrate in outline when viewed from the distal aspect. Its antero-posterior diameter is 4 mm. Anteriorly the articular area ceases at the edge of the shallow trochlear fossa; posteriorly it ends at the edge of the deeper olecranon fossa. The capitellar surface for the head of the radius is proportionately large. It forms a segment of a short, squat cone whose base lies adjacent to the lateral epicondyle and truncated apex abutting on the lateral surface of the trochlear portion of the articular area. A pulley-like groove is thus formed between trochlea and capitellum, deeper anteriorly, shallower behind and below.

The olecranon is elongated axially, narrow transversely and much excavated on its articular face. Its articular surface is smooth, ungrooved. Its widest part is proximally, where it attains a width of 4 to 5 mm. A small sigmoid notch, covered with cartilage continuous with that on the main part of the bone, lies laterally near the volar end of the main articular area. This receives the head of the radius, held in position by the broad orbicular ligament. A triangular pad of fat occupies the angle dorsally between the lateral edge of the olecranon and the head of the radius.

A capsular ligament scarcely exists, being reduced front and back to areolar tissue beneath the muscles and tendons crossing the joint. Medially and laterally, however, there are strong collateral ligaments passing from the non-articular part of the humeral condyles obliquely to the corresponding non-articular bone on the olecranon process. That on the radial side is more transversely disposed; while the medial collateral band passes more to the apex of the olecranon. On the lateral side also there is a distinct band passing from the forepart of the olecranon to the lateral part of the orbicular ligament of the radius.

The orbicular or annular ligament forms about three-fourths of a circle. It is attached by its ends to the anterior and posterior margins of the lesser sigmoid notch. The ligament is more constricted, though thinner distally, narrowing on to the neck of the radius, thus serving to prevent downward dislocation.

The interosseous membrane between radius and ulna is very tenuous, with the fibres arranged as in the human structure. It is reinforced proximally by a stratum of fibres in the opposite direction corresponding to the oblique cord of anthropotomy, but more membranous in character.

8. WRIST JOINT

As in *Tarsius*, there are two synovial cavities separated by a vertical fibrous septum passing from the lower end on the inferior radio-ulnar union to the interosseous ligament connecting the semilunar with the cuneiform bones of the carpus. There is thus a radio-carpal joint on the preaxial side and a combined cubito-cuneiform-pisiform joint postaxially, there being no triangular fibro-cartilage between the ulna and cuneiform. The ulnar articular facet has a downwards and forwards aspect and is contributed to as much by the short styloid process as by the head of the bone. There is a strong volar ligament connecting the ulna and cuneiform. The facet on the base of the pisiform is dual in nature, the two parts being separated by a transverse ridge; the proximal facet is apposed to the ulna and the distal to the cuneiform.

9. CARPAL ARTICULATIONS

The proximal row of carpals is held together at the mid carpal level by a strong interosseous ligament between scaphoid and semilunar and a volar ligament (without extension between the bones) connecting semilunar and cuneiform. Proximally interosseous ligaments occur between all three bones. The pisiform does not enter into the mid-carpal joint, the proximal surface of which is formed solely by scaphoid, semilunar, and cuneiform. The distal aspects of scaphoid and semilunar are of approximately triangular outline, whereas that of the cuneiform is rounded, almost circular.

10. SACRO-ILIAC JOINT

The auricular facets on sacrum and ilium describe a short, boomerang-shaped contour, with the concavity located dorsally. These are covered with a thin layer of articular cartilage, but no synovial layer was present. The two bones are connected around the edges of the auricular facet by thickened periosteum; the ventral periosteum being especially thick and membranous, while the dorsal connection is more ligamentous, occupying the whole of the afore-mentioned concavity.

11. HIP JOINT

The capsule is a cylindrical sleevelike structure extending from the margins of the glenoid labrum of the acetabulum to the distal part of the neck of the femur. Unequal in thickness, its strongest region is below where the ischio-femoral band is well differentiated, and directed more obliquely than the fibers of the major part of the capsule. The capsule is lined within by synovial membrane which is modified above and below near the femoral head to form short, triangular superior and inferior retinacula. The superior retinaculum is slightly ventrad to the center of the uppermost part of the capsule. Articular cartilage is confined to the head of the femur and, in the acetabulum, to a horseshoe-shaped peripheral zone averaging 2.0 mm. wide, but broadening at its dorsal extremity and narrowing ventrally. The nonarticular area is occupied by fat, synovial membrane and the attachment of the stout ligamentum teres. The latter is rounded, not flattened as in Man, but very strong. Its femoral attachment occupies the whole of the fovea capitis. The ligament is tense when the femur is slightly flexed, laterally rotated and abducted. Its general direction is from above downwards, i.e., ventrad.

The peripheral margin of the articular cartilage on the femoral head is irregular. It extends distally to the greatest degree on the dorsal side, and above, where it presents a convex contour. Inferiorly the edge is concave. There is a further concavity on the ventral side separated by a slight convexity from the deeper inferior recession.

12. KNEE JOINT

This is the largest of the diarthroidal joints, and involves the femur, patella, and tibia. The patella is associated proximally with an accessory suprapatella similar to that occurring in *Hapale* and *Tamarin* (Jamieson, 1904; Retterer and Vallois, 1912).

The femoral articulating area describes, when viewed from below, an oblong contour, with the long axis in the sagittal plane. The joint is therefore longer antero-posteriorly than in the transverse dimension compared with that of Man. Moreover, the lateral condyle is more elongated than the medial, being especially prominent anteriorly. The axes of the two condylar areas diverge somewhat posteriorly. The intercondylar notch separates them over slightly more than one-third the antero-posterior extent of the articular area.

The articular area for the patella is therefore relatively extensive on the distal surface of the femur.

a. CAPSULE

The capsule of the joint is of the usual imperfect construction, being deficient anteriorly where it is replaced by the patella, suprapatella, and ligamentum patellae. It is reinforced laterally by the ilio-tibial band of fascia lata and by fibrous derivatives from the vasti. On the two sides the capsule proper is attached to the sides of the condyles 1.0 mm. above the articular margin. On the lateral condyle the attachment skirts around the fossa to which the tendon of origin of the popliteus is connected, thus including the tendon within the joint. Posteriorly it covers the backs of the condyles and attaches to the bone immediately above them and to the intercondylar line at the upper limit of the intercondylar notch.

At each supracondylar attachment a deficiency occurs owing to the incorporation within the capsule of the sesamoid bones (fabellae) developed in the two heads of the gastrocnemius. Each fabella has an articular cartilage-covered facet on its deep surface applied to the condylar articular cartilage. The lateral fabella is slightly the larger with a transverse diameter of 2.7 mm. compared with 2.4 mm. for the medial. The articular facet on each is approximately circular in outline with a slight flattening along the proximal edge.

The tibial attachment of the capsule is to the sides of the condyles some 2.0 mm. distal to the upper margin, to the lower margin of the intercondylar notch posteriorly, and to the front along lines extending from near the articular margins at the sides running forwards and downwards to the sides of the tubercle. At the back of the lateral condyle the capsule is deficient where it is pierced by the popliteus tendon and the bursa accompanying same.

The posterior aspect of the capsule receives no reinforcement from the semimembranosus tendon. Its lower part, however, is strengthened by overlying transverse fibres of the uppermost part of the popliteus muscle.

Collateral ligaments are present on the two sides of the joint, serving to strengthen the capsule, though not incorporated in it. The lateral one, some 9 mm. long, connects the upper part of the lateral aspect of the femoral condyle with the head of the fibula. The fibular attachment is not to the summit of the head, but some distance down on its lateral surface. The medial collateral ligament, somewhat more bandlike, is 10 mm. long and connects the medial surface of the medial femoral condyle with the medial surface of the tibial tubercle; it is therefore somewhat obliquely disposed. Its great length is due to the proximo-distal elongation of the tibial tubercle.

b. SEMILUNAR CARTILAGES

These are two curved fibro-cartilages applied to the articular surfaces on the corresponding tibial condyles. Each is attached firmly by (a) a peripheral ligamentous tract (coronary ligament) to the edge of the corresponding tibial condyle and (b) by ligamentous connections between the anterior and posterior horns of each cartilage and a point in the intercondylar area.

In spite of the greater antero-posterior extent of the lateral femoral condyle, the semilunar cartilages are shaped and their attachments arranged as in the human knee joint. Thus the lateral cartilage is C-shaped and the medial more oval in outline, with its two horns tending to embrace those of the lateral meniscus. The two horns of the lateral cartilage are attached close together, the anterior one to the back of the eminence. The popliteus tendon disturbs the marginal connection with the tibia, and a sesamoid bone (cyamella) is developed within the tendon at this site.

The anterior horn of the mesial meniscus is attached to the forepart of the intercondylar area in front of the anterior cruciate ligament, and its posterior horn to the posterior part of the area in front of the attachment of the posterior cruciate ligament. Its coronary ligament is uninterrupted and fused with the capsule.

c. INTERCONDYLAR SEPTUM

This consists of a complex of structures within the articular cavity of the joint dividing it imperfectly in a sagittal plane into two chambers. The septum extends from the non-articular parts of the intercondylar notch of the femur to the non-articular area on the upper surface of the tibia. It includes the two cruciate ligaments, which together comprise the posterior part of the septum, and, in front of these, the infrapatellar fold. Its apex is attached to the most anterior part of the intercondylar notch on the femur; its base extends from the area below the edge of the articular-cartilage-covered area of the patella to the most anterior part of the intercondylar area on the tibia and the area below that on the upper part of the ridge leading to the tubercle. From the sides of its base alar folds, also containing fat, are developed as fringes, prolonged each side some 4 mm.

The cruciate ligaments call for no remark as their positions and attachments differ in no material respect from those of Man.

13. ANKLE JOINT

A hinge-joint with extensive synovial cavity, which does not, however, extend upwards between tibia and fibula. The inferior tibio-fibular articulation is a pure syndesmosis, with the transverse ligamentous connections continuous inferiorly with the capsular ligament of the ankle joint.

The capsular ligament is attached to the articular margins around both tibia and fibula and, as in other hinge-joints, is thin anteriorly and behind, but greatly thickened on the two sides. On the lateral side the chief bond of union is a horizontally disposed bundle connecting the deep surface of the lateral malleolus, beyond its articular area, with a rough depression on the lateral aspect of the talus in the concavity distal to the comma-shaped articular surface. The medial or deltoid ligament similarly connects the tip of the medial malleolus with the medial surface of the body of the talus distal to the articular facet. Reinforcement by talo-calcaneal bands is minimal, additional strength being supplied rather by the various tendons passing over the capsule.

14. PEDAL JOINTS

a. TALO-CALCANEAL ARTICULATIONS

These are two in number, posterior and anterior. The former is a gliding joint between the opposed posterior facets on the two bones (see p. 31 above). It is surrounded by a capsule which is attached to each bone just beyond the edge of the articular cartilage. The capsule is thin and fairly lax everywhere, except anteriorly where it is combined with the capsule of the anterior joint to form the strong interosseous ligament which occupies the sinus tarsi, attaching to the under surface of the neck of the talus above and below to the floor of the sulcus tarsi of the calcaneus.

The anterior joint is formed between the head of the talus and the sustentaculum tali; it is shared with the talar facet on the navicular and by the strong plantar "spring" (inferior calcaneo-navicular) ligament extending from the under surface of the sustentaculum tali to the whole of the inferior surface of the navicular. This joint is virtually a ball-and-socket joint with movement limited by the short dorsal parts of the capsule to gliding and rotation. The interosseous ligament forms the posterior wall of the capsule.

Dorsally, support is afforded by a bifurcated ligamentous band (1.45 mm. broad) extending from the neck of the talus forwards to end in two slips, one to each of the second and third metatarsals. Lateral to this and on a slightly deeper plane is another bifid band homologous with the ligamentum bifurcatum of human anatomy. This arises on the superior aspect of the calcaneus lateral to the anterior talar facet. Distally it splits into two bands, one to the dorsal surface of the cuboid and the other ending on the upper and medial surfaces of the navicular, where it blends below with the spring ligament thereby completing the capsule.

b. MID-TARSAL JOINT

This is comprised by the anterior talo-calcaneal (talo-calcaneo-navicular) joint just described, combined with the calcaneo-cuboid articulation. In the latter joint a capsule unites the two bones and the articular cavity is independent of the other part of the mid-tarsal joint. The capsule is supplemented dorsally by the lateral limb of the ligamentum bifurcatum deep to which lies the dorsal calcaneo-cuboid ligament—a part of the capsule. On the plantar surface the plantar calcaneo-cuboid ligament connects the inferior edge of the cuboid facet of the calcaneus with the inferior aspect of the cuboid. Superficially this is supplemented by the long plantar ligament, a dense tract of fibres 2.7 mm. in breadth. This commences on a tubercle or ridge on the forepart of the inferior surface of the calcaneus and extends forwards over the whole breadth of the cuboid, some of its fibres attaching thereto, but others proceeding forwards to end on the plantar aspect of the bases of the third, fourth, and fifth metatarsals.

Movements at the mid-tarsal joint involve dorsiflexion at the calcaneo-cuboid articulation combined with medial rotation; this is combined with medial rotation of the head of the talus in its socket formed by the navicular, spring ligament and sustentaculum tali. The combined effect is inversion of the foot.

c. CUNEO-NAVICULAR ARTICULATIONS

The shapes of the bony articular surfaces involved here are considered on p. 32. The bones are held together by a common, continuous capsular ligament attached near the articular margins, and the synovial cavity is continuous with that of the cubo-navicular joint.

d. CUBO-NAVICULAR JOINT

The two bones are united by thin dorsal and thick plantar ligaments. There is also a strong interosseous ligament towards the plantar side of the articular facets.

e. CUBO-CUNEIFORM JOINT

This involves a synovial cavity placed distal to that of the preceding joint. A thick interosseous ligament joins the bones and is an inward extension from the thick plantar part of the capsule.

f. INTERCUNEIFORM JOINTS

The articular areas have already been described (p. 32). Two dorsal ligaments connect the three bones. A strong plantar intercuneiform ligament connects the lateral side of the base of the entocuneiform with the apex of the mesocuneiform. There are also interosseous ligaments connecting both sides of the mesocuneiform with its neighbors. That on the medial side is a rounded bundle attached to the middle of the bones; whereas the lateral one is more diffuse and the articular cartilage-covered area confined to two triangular surfaces at the anterior and posterior dorsal angles of the lateral surface of the bone.

g. TARSO-METATARSAL AND INTERMETATARSAL JOINTS

The base of the first metatarsal articulates with the entocuneiform only. The second articulates mainly with the mesocuneiform, but also with ento- and ectocuneiforms. The fourth metatarsal articulates solely with the ectocuneiform. The remaining two metatarsals are united to the cuboid.

Intermetatarsal joints occur between the heads of neighboring metatarsals except between I and II. These are diarthrodial joints whose synovial cavities communicate with the neighboring intertarsal cavities.

MYOLOGY

I. TELA SUBCUTANEA

Subcutaneous areolar tissue shows, locally, deposits of normal fat. This is concentrated more particularly in certain locations such as the axilla, root of the neck and inguinal region. It also occurs in intermuscular planes and intervals and in the orbit, as well as around the kidneys and between the laminae of the mesentery.

In certain locations the fat is either wholly or partly of the "brown" or "glandular" variety as typified by that of the so-called hibernating gland of the hedgehog and many rodents. In *Callimico*, brown fat has been observed at the undermentioned sites: (*a*) the anterior part of the temporal fossa; (*b*) in the suboccipital triangle; (*c*) in the pterygoid region; (*d*) the deeper parts of the axilla, deep to normal fat; (*e*) deep to the rhomboideus muscle (see fig. 22).

In the hinder part of the lumbar region between the dorsal and middle layers of the lumbar fascia ordinary pale fat was abundant.

II. DEEP FASCIAE

These are developed as in Man and other Primates and call for no special comment of a general nature.

III. THE VOLUNTARY MUSCLES

1. FACIAL MUSCLES

As the skin was required by the American Museum of Natural History, no opportunity occurred for dissection of the facial muscles. The only muscle of this group remaining on the head after removal of the pelt was the large retrahens aurem (auricularis posterior). This is a single elongated flattened fleshy slip arising over the extreme posterior limit of the cranium and inserting on the cranial aspect of the pinna. This accords with Ruge's (1887) finding in *Hapale penicillata*, except that the band is both relatively and absolutely broader. Huber (1931) shows several slips in *Oedipomidas*. Lightoller (1934) also found, contrary to Beattie, that in *Hapale* the muscle is a single belly taking origin near the inion, inserting into the posterior surface of the auricular cartilage between the slips of the cervico-auriculo-occipitalis.

The following muscles were defined on the left side of the head of the female PP 119. Nomenclature follows that of Lightoller (1934).

1. *Platysma*

Largest of the facial muscles, it has a very broad origin reaching back to the dorsal median line of the neck. As in *Hapale*, it may be divided topographically into notoplatysma and trachelo-platysma, the tip of the acromion being the point of separation. It resembles that described by Ruge (1887) for *Hapale*, but as Lightoller found, it lacks all dermal attachments.

Below the cartilage of the pinna the cranial border of the platysma receives the fibres of the m. cervicalis transversus. Fibres from the shoulder region and those of the trachelo-platysma which spring from the chest wall behind the clavicle are noticeably paler than the remainder; all the fibres converge to form a flat band

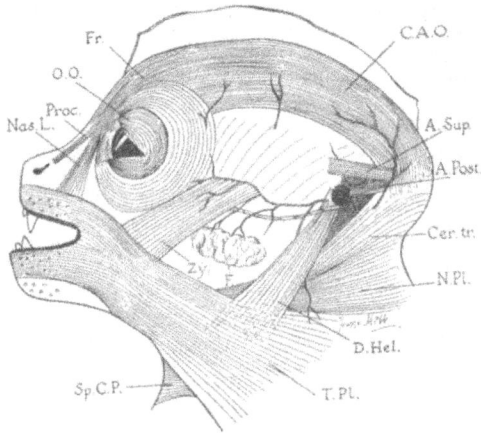

FIG. 13. Dissection of the facial muscles of the left side.

some 12.5 mm. broad at the junction of the head and neck and over the angular region of the mandible. Thence they proceed towards the angle of the gape. A few spread over the cheek, but most enter the lips forming the superficial layer of musculature therein. Those to the upper lip are reinforced by fibres of the auriculo-labialis.

2. Orbicularis Oculi

Contrary to Lightoller's findings in *Hapale,* there is a distinct macroscopic differentiation between the palpebral and the orbital fibres of this muscle. The former is a much thinner and paler stratum, the latter appreciably thick and ruddy compared with most of the facial muscles. Palpebral fibres appear, as in *Hapale,* to have no bony attachment; the orbital fibres gain indirect attachment through the intermediacy of the medial palpebral ligament. The most supero-medial fibres connect with those of the naso-labialis and procerus and lateral to these some interlacement with frontalis fibres occurs. The most inferior fibres enter the upper lip.

3. Temporo-labialis

No trace of a temporo-labialis was found. It is to some extent replaced by a thin feebly fleshy lamina which Lightoller refers to in *Hapale* as temporo-auriculo-labialis, taking origin from temporal fascia anterior to the ear and sweeping forwards to the upper lip. *Callimico,* however, shows distinct advance on *Hapale* in so far as, deep to the above, is a well-defined zygomaticus with broad origin from the malar bone forming a parallel-sided muscle (4.5 mm. across) which passes obliquely forwards and downwards to the upper lip just medial to the angle of the gape, completely covering the orbicularis oris.

4. M. Maxillo-naso-labialis

This occurs as in *Hapale,* taking origin from the maxilla close to the alveolar border above the two hinder premolars.

5. M. Naso-labialis

This is a bundle continuous above with the medial fibres of frontalis. They pass medial to the medial palpebral canthus over the root of the nose, then expanding in a fan-shaped manner the more medial or nasal fibres inserting on the external nose near the narial margin; lateral (labial) fibres proceed to the upper lip, entering between the zygomaticus and the orbicular strata.

6. M. Procerus

This is present as in *Hapale.* It lies still more medially than the preceding in contact with its fellow in the median line. It is well marked and partly overlaps the medial fibres of the naso-labialis. It inserts in the tissue occupying the depression between the two external nares.

7. Corrugator Supercilii

This is about equally developed as in *Hapale* arising from the medial part of the supraorbital margin and spreading obliquely lateral at first deeply to orbicularis oculi, its fibres insinuate between those of neighboring muscles to end on the deep surface of the skin of the brows, especially about the follicles of the supraorbital sinus-hairs.

8. M. Frontalis

This forms an extensive sheet over the whole frontal region from the brows to the vertex where, as in *Hapale,* it becomes contiguous with fibres of the cervico-auriculo-occipitalis, the two muscles together forming a complete fleshy covering over the whole scalp. Laterally there is indistinct differentiation from the orbito-auricularis, but the latter is more attenuated and more fascial in nature than in *Hapale.* The latter muscle or fascial sheet has an anterior attachment to the lateral orbital margin. Medial frontalis fibres are continued into the procerus and naso-labialis; lateral fibres perforate the orbicularis oculi to insert in the skin of the brows.

9. Sphincter Colli Profundus

The very delicate anterior portion referred to by Lightoller in *Hapale* is apparently lacking. The posterior portion, however, is well developed, consisting of oblique fibres deep to the trachelo-platysma. Fibres from the two sides interlace freely in the mid-ventral line of the neck; caudally fibres lie superficial to the sternum. The posterior fibres sweep upwards and forwards obliquely deep to platysma and end on the cheek anterior to the ear, where they interlace or become continuous with downwardly directed fibres of the depressor helicis.

10. *Orbicularis Oris, etc.*

As in *Hapale* the upper lip has no pars marginalis and the m. caninus, apart from its connection with the bone over the root of the canine, is not separable from the orbicularis oris (pars peripheralis). At the angle of the gape is a very stout modiolar bundle largely formed from fibres derived from the buccinator, but with some superficial contribution from trachelo-platysma; but there are also some superficial fibres derived from the canino-orbicularis complex entering at right angles to the buccinator fibres. Medially the pars peripheralis in the upper lip is inseparable without artificial aid from the mass of fibres parallel with those of orbicularis but inserting on the lateral margin of the external nares superficial to the descending fibres of the naso-labialis. This is the m. maxillo-naso-labialis. A small marginal bundle is present in the lower lip forming its free margin in the same plane as the pars peripheralis. Towards the modiolus the pars marginalis passes deep to the pars peripheralis and interlaces with the lower fibres of buccinator.

Fibres from the bony edge of the apertura pyriformis and adjacent nasal cartilage proceed downwards to merge with fibres of the pars peripheralis of the upper lip. These have been taken to represent a m. incisivus superior. There is no corresponding structure in the lower lip.

11. *M. Mentalis*

This forms a hammock-like structure supporting the inferior labial part of the orbicularis oris, some fibres

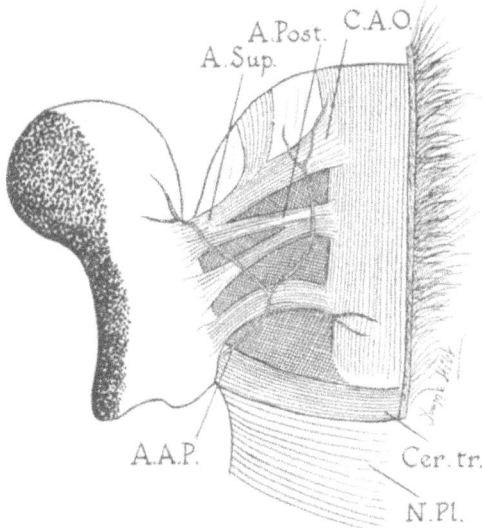

FIG. 14.   Dissection of the facial muscles of the left side from behind.

being superficial and others deep thereto. Its fibres arise from the alveolar margin just distal to the gum over a line extending from the median plane to the level of the lateral limit of the root of the canine. Fibres pass medially to insert with the skin, the deeper ones posterior to the orbicularis and the others superficial thereto and somewhat inferiorly. Fibres of opposite sides interlace over the chin area.

12. *Extrinsic Muscles of the Auricle*

Auricularis posterior arises from the mid-line near the inion and at first constitutes a single fleshy belly. In the present specimen, however, some fasciculation occurs as the fibres pass on to the cranial surface of the auricular cartilage. The fascicles proceed between the auricular slips of the more superficial cervico-auriculo-occipitalis.

Auricularis superior is a distinct separate slip arising from fascia near the lateral border of cervico-auricular-occipitalis as the latter proceeds over the cranium opposite the auditory meatus. Its belly narrows to a single slip which inserts on the cranial aspect of the auricular cartilage some 5 mm. superior to its root.

As Lightoller found in all the lower Primates he examined there is no true auricularis anterior which is replaced functionally by the powerful orbito-auricularis.

2. OCULAR MUSCLES

The four recti differ from the other orbital muscles in their deeper color. All are relatively short, but bulky muscles arising by tendon from the fibrous ring around the optic foramen; but the lateral rectus has the usual two roots of which the upper is tendinous and the lower fleshy. The muscles swell up to form fusiform fleshy bellies, but, as they make contact with the globe, they broaden, flatten and become extremely attenuated, finally giving rise to resplendent aponeuroses, with the constituent fibres fanning out to their insertions. The superior rectus inserts some 9.1 mm. posterior to the limbus, the medial rectus 7.9 mm. therefrom and the lateral 6.8 mm. and inferior 4.5 mm. The distance between superior and lateral recti is less than between superior and medial. In the interval between the two latter the superior oblique is located.

The superior oblique is the longest and narrowest of the orbital muscles. Arising from the upper and medial side of the fibrous annulus, it proceeds forwards between the levator palpebrae superioris and the upper edge of the medial rectus. Becoming tendinous, it passes through a fibrous pulley in the usual way, turning thence laterally and backwards to insert on the forepart of the globe just behind the medial limit of the conjunctival insertion of the levator.

The inferior oblique is a broadish parallel-sided fleshy band arising from the maxilla as it forms part of the orbital floor. It proceeds laterally and posteriorly on the surface of the globe to which it is connected by

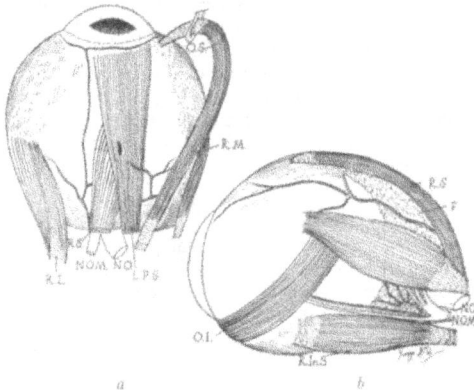

FIG. 15.   Left ocular muscles (*a*) from above, (*b*) oblique view
from the lateral side and slightly from above.

areolar tissue. It passes beneath the lower border of the
lateral rectus and inserts deep thereto by a broad
aponeurosis.

Levator palpebrae superioris is a very pale muscle,
narrow behind and broad anteriorly. Arising from the
uppermost part of the annulus, its fibres become fleshy
and proceed forwards dorsal to the superior oblique.
The fibres radiate somewhat towards their insertion
which is chiefly into the eyelid, but the deepest fibres
find attachment to the fornix of the conjunctiva some
3.8 mm. from the limbus.

Ottley (1879) pointed out certain differences between the Hapalidae and Cebidae in the matter of
detailed insertion of the ocular muscles. Whereas in the
Cebidae the superior rectus and superior oblique are
inserted almost at right angles to each other (completely so in *Cebus*), in *Hapale penicillata* and *Leontocebus rosalia* the rectus is so obliquely inserted as to
approach the direction of insertion of the superior oblique. Another feature is that the line of attachment
of the lateral rectus is convex forwards, recalling the
condition in lemurs.

In *Callimico* I find the superior rectus insertion rather
oblique, but not as oblique as that of the superior oblique, the two lines meeting at an angle of approximately
45°, the latter being nearer the limbus. The lateral
rectus insertion is linear. As in the Hapalidae, there
is no choanoid slip.

3. MUSCLES OF MASTICATION (figs. 16, 58)

The masseter consists of superficial and deep portions, well differentiated from each other. The bulky
superficial part arises mainly by a short tendon, concentrated on the forepart of the superficial aspect of the
zygoma. Tendinous fibres radiate over the superficial
aspect of the muscle, but deeply are replaced by fleshy
fibres taking the same directions. Uppermost fibres

run almost horizontally backwards, the fibres becoming
increasingly oblique from above downwards; the most
anterior fibres are still oblique though nearer the
vertical than the horizontal; i.e., the anterior edge of
the muscle is not vertical. None of the deep masseter
is visible until the superficial mass is reflected. The
superficial layer inserts along the lower edge of the
mandible in its posterior one-third, and around the
broad angular region as far as its most posterior upturned point. The insertion overlaps slightly on to
the superficial aspect of the bone, but it is widely
separated from the insertion of the deep masseter.

The deep masseter is a wholly fleshy mass arising
from the hinder three-quarters of the lower border of
the zygoma and the whole of its deep aspect. Its fibres
descend almost vertically and insert over the lateral
aspect of the vertical ramus of the mandible as far as
the level of the upturned process on the angular region,
but is separated from it by a wide space. The anterior
border becomes tendinous, the insertion of this tendon
being responsible for a curved bony ridge descending
from the point of union of vertical and horizontal rami
downwards and backwards for 6 mm. (see p. 24 above).
To some extent the anterior tendinous fibres blend
deeply with the fibres of the temporalis as they insert
on the coronoid process and anterior border of the
vertical ramus, as Beattie found in *Hapale*. In almost
every detail the masseter of *Callimico* agrees with that
given by Starck (1933) for *Hapale* and *Leontocebus*.

Temporalis

As in Hapalidae this is covered by a very strong
temporal fascia which gives origin to fleshy fibres on
its deep surface. The fascia is attached to the temporal
line which, commencing behind the upper limit of the
orbit, sweeps backwards parallel with the vault, but
some 9 mm. from it. The muscle is therefore less
voluminous than in *Leontocebus*, but resembles rather

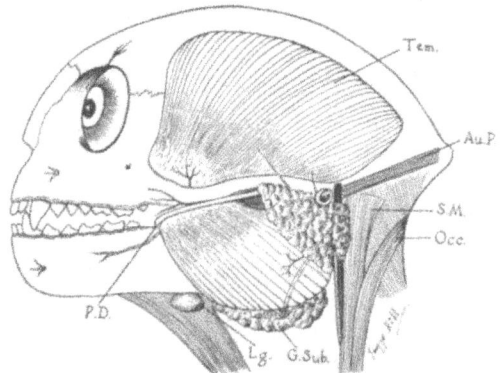

FIG. 16.   Dissection of left side of head to show
muscles of mastication, etc.

that of *Hapale*. Posteriorly the limit of the muscle falls short of the inion, the temporal line turning sharply downwards and then forwards towards the suprameatal region. The muscle is thus less extensive posteriorly than in *Leontocebus*, but more so than in *Hapale*, where the hinder limit is more evenly rounded and less parabolic in outline.

Near the temporal line the fleshy mass is extremely thin (0.8 mm. thick) but rapidly thickens up to 1.3 mm. in the center. Anterior fibres are almost vertically disposed and fall short of the orbital margin by some 5 mm. These fibres arise from the temporal aspect of the bulbous lateral wall of the orbit. Fibres adopt increasingly oblique positions in the more posterior parts of the muscle, at the same time increasing in length. The longest fibres spring from the apex of the parabola of the temporal ridge. Below that the fibres are shorter and almost horizontally disposed. All these fibres converge to a strong tendon which inserts on the tip of the coronoid process, its anterior margin and the anterior border of the ascending ramus of the mandible, and over a considerable surface of the deep aspect of the bone along the same distance (see fig. 8). The above-mentioned postero-inferior fibres are partly differentiated from the remainder, inserting on a deeper plane; they are partly overlapped from above by the more dorsally placed fibres. They sooner give rise to tendinous fibres, this tendon also receiving the deeper fibres from the infratemporal fossa and likewise those from the deep aspect of the zygoma. The insertion of the temporalis is not wholly tendinous, there being some fleshy attachment posterior to the main tendon, these fleshy portions being incompletely differentiated from the deep portion of the masseter.

Pterygoideus Lateralis

A small dark-fibred fleshy muscle with fibres disposed mainly horizontally. Its origin is from the lateral aspect of the lateral pterygoid plate and the orbital plate of the malar. Its fibres converge to a thick tendon which attaches principally to the front edge of the meniscus of the temporo-mandibular joint, with some fibres connecting with the neck of the mandible and the anterior part of the capsular ligament.

Pterygoideus Medialis

Much larger than the foregoing, but similar in its dark fleshy fibres, it has a quadrate outline, but with the lower border somewhat irregular. Its fibres are parallel and directed almost vertically downwards, with only a slight caudal trend. Arising from the meso-pterygoid fossa, and closely related posteriorly to the front wall of the bulla and to the thickened stylo-mandibular ligament, it inserts by fleshy attachment, to an area on the deep aspect of the mandible ventral to the inferior dental foramen, extending thence backwards and downwards almost to the angle. It does not gain the inferior border of the mandible.

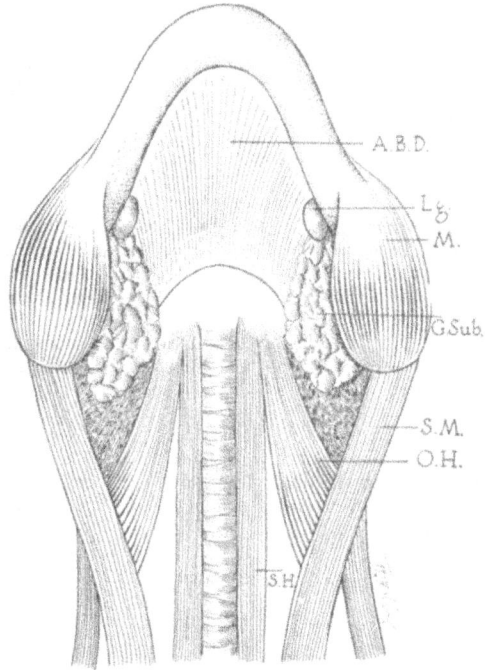

Fig. 17. Dissection of suprahyoid and infrahyoid muscles.

4. SUPRAHYOID MUSCLES (fig. 17)

Digastricus

The posterior belly is conical, elongated and mainly fleshy. It arises on the mastoid region deep to the insertion of sterno-mastoid and trachelo-mastoid. Its belly proceeds forwards and downwards, narrowing rapidly to a thin, elongated but strong intermediate tendon. From the tendon a thin fleshy sheet continues the muscle forwards on the interramal region where it forms, with its fellow, a complete covering to the deeper mylohyoid. These anterior bellies consist of parallel fibres, the medial ones forming a very attenuated sheet, whereas the lateral ones form a thicker stratum. Posteriorly a fibrous arch constitutes a caudal edge to the muscle, passing from the intermediate tendon of one side to that of the other. This arcade is attached to the body of the hyoid by aponeurotic fibres. The insertion of the anterior belly is along the whole length of the medial surface of the mandible, near its lower border, from the anterior limit of the masseteric insertion forwards to the symphysis.

In *Hapale* the two anterior bellies are unconnected but Windle describes a fibrous arch between the intermediate tendons in *Leontocebus*.

Mylohyoideus

This forms a complete fleshy diaphragm across the mandibular rami deep to the preceding, but the fibres proceed mainly transversely instead of horizontally. Only the fibres of the caudal one-third of the muscle are connected to the hyoid and this through the same aponeurosis as that connected with the anterior digastric bellies. The remaining fibres end in a median raphe from which a fibrous septum proceeds upwards, between the genio-hyoidei, and genio-hyo-glossi, into the body of the tongue. *Genio-hyoidei* and *genio-hyo-glossi* are well differentiated from each other, and there is a layer of fat between them, especially posteriorly, near the hyoid. Both have firm hyoid attachments, and both are completely fleshy. Beattie states that in *Hapale* the genio-hyo-glossus is not differentiated from the hyoglossus.

Hyoglossus

This is quite distinct, proceeding forwards and upwards on each side from the greater hyoid cornu into the tongue, with the usual relations to the local nerves and vessels.

5. INFRAHYOID MUSCLES (fig. 17)

Sterno-hyoideus

This is an extremely thin band of pale fibres 3.8 mm. broad, almost touching its fellow in the mid-line. It arises from the dorsum of the manubrium sterni, passing forwards upon the trachea and sterno-hyoid muscle to insert on the ventral aspect of the hyoid body in line with the insertion of omohyoideus.

Omohyoideus

This is a similar but broader (4.5 mm.) band coursing obliquely from the shoulder region to the hyoid. Its origin is from the ossified suprascapular ligament. Below it almost touches the ventral border of the trachelo-acromialis to which it is connected by fascia. Above it approaches the sterno-hyoideus to which it is also connected by a fascial sheet. Its insertion is by fleshy fibres upon the greater cornu in line with that of sterno-hyoideus.

Sterno-thyroideus

A narrow band (only 3.1 mm. broad) but composed of darker fibres forming a denser mass than sterno-hyoideus. It arises on the manubrium sterni posterior to the origin of the more superficial muscle and inserts upon the oblique line of the thyroid ala.

Thyro-hyoideus

A small, parallel-sided pale muscle arising from a distinct oblique ridge on the thyroid ala with its fibres proceeding cranially and medially to insert on the body of the hyoid deep to the preceding.

6. MUSCLES OF THE NECK

Sterno-cleido-mastoideus

As in *Hapale*, this consists of two parts, a superficial sterno-mastoideus arising from the sternum and capsule of sterno-clavicular joint and a deeper stratum (cleido-occipitalis) separate at its origin from the medial end of the clavicle, but joining the deep aspect of the main mass, which constitutes an oblique band coursing up the side of the neck to end in the mastoid region and nuchal crest. The posterior border gives off an occipital slip, some 2 mm. broad, to the dorsal part of the nuchal crest.

Trapezius and rhomboideus capitis are considered below with the appendicular muscles.

Splenius

This is an obliquely disposed flat fleshy sheet lying deep to trapezius and rhomboideus capitis, and partly superficial to complexus (longissimus capitis). It arises by slips from the spines of the first three thoracic vertebrae and inserts on the lateral part of the nuchal line deep to cleido-occipitalis. At its insertion the muscle is 9 mm. broad.

A couple of small slips leave the lateral border of the muscle to insert on the transverse processes of the first and second cervical vertebrae, constituting a splenius colli, as Beattie found in the marmoset.

Complexus (Longissimus capitis)

This is a thick fleshy mass for the most part parallel sided and composed of axially running fibres. It arises by slips from the transverse processes of thoracic vertebrae 2–4. At first the lamina formed by their union is obliquely disposed and placed deep to the origin of splenius and trachelo-mastoideus. As the latter muscle passes off it, the complexus becomes more horizontally disposed with dorsal and ventral aspects, the latter separated deeply from the short suboccipital muscles by a very dense fascial stratum.

This stratum meets its fellow in the median line and there becomes even more dense, constituting a ligamentum nuchae. This contains no demonstrable elastic tissue, but is 0.5 mm. broad. The complexus is inserted upon the nuchal plane, just beneath the medial part of the nuchal crest. At its insertion the muscle is 7 mm. broad.

Trachelo-mastoideus

This is a long narrow fleshy band, some 4 mm. broad, arising by four slips from the transverse processes of the last cervical and first three thoracic vertebrae. It inserts over the mastoid region superficial to the origin of the posterior belly of the digastric. Its lowest slips are covered superficially by the serratus dorsalis anterior.

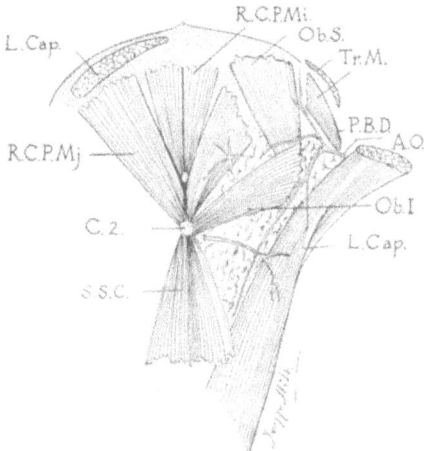

FIG. 18.   Dissection of suboccipital region.

7. SUBOCCIPITAL MUSCLES

Beattie describes these in *Hapale* as small and weak, but in *Callimico* they are all robust, fleshy muscles of triangular outline. They are disposed in an almost horizontal plane in view of the position of the occipital squama, the major part of which is occupied with the insertions of the two posterior recti and the superior oblique (fig. 18). The suboccipital triangle is reduced in correlation with the large area of the muscles bounding it and it is filled with some small lobules of very dark fat composed partly of glandular or "brown" fat and partly of ordinary adipose tissue.

Rectus Capitis Posterior Major

This arises by musculo-tendinous fibres from the spine of the axis. It broadens to a fan-shaped belly, the fibres proceeding obliquely craniad to insert over a large area of the occipital squame antero-medial to the insertion of the superior oblique. Its lower border, laterally, is related to the suboccipital nerve.

Rectus Capitis Posterior Minor

This is similar to the preceding, but on a slightly deeper plane, and more sagittally directed. Its origin is musculo-tendinous, from the posterior tubercle of the atlas. Only its extreme lateral edge is overlapped by the larger muscle. It inserts on the occipital squame more medially and slightly nearer the lambda than the rectus capitis posterior major.

Obliquus Capitis Superior

This is even more robust than the larger posterior rectus. Its thick musculo-tendinous origin is from the lateral mass of the atlas. Its fibres fan out and have a general craniad and medial direction. Insertion is

by fleshy attachment to the lateral part of the occipital squame over a large area, deep to complexus and trachelo-mastoid insertions and medial to the origin of the posterior belly of the digastric.

Obliquus Capitis Inferior

This springs by a robust musculo-tendinous cord from the lateral part of the spine of the axis. Its fibres spread upwards and laterally to a wide area of insertion on the lateral mass of the atlas, partly hiding the origin of the preceding muscle.

Semispinalis Colli

A fleshy mass clothing the arches of all the lower cervical vertebrae. At its cranial end it narrows abruptly to terminate in a musculo-tendinous mass which concentrates upon the posterior surface of the spine of the axis.

8. PREVERTEBRAL MUSCLES (fig. 19)

Rectus Capitis Anterior Major

This is a powerful fleshy mass clothing the lateral parts of the ventral surface of the cervical vertebrae, being medially in contact with the longus colli, which clothes their centra. It arises by tendinous slips which rapidly become fleshy, chiefly from the ventral tubercles of the transverse processes of C.2, C.3, and C.5, there being no slip from C.4. Dorsally the muscle is reinforced by fibres coming from more posterior vertebrae (C.4–C.6). These join the slip from C.5, but some almost immediately leave the slip on its medial side dorsally to gain attachment to C.2 dorsal to the origin of the uppermost slip of the main muscle mass. A broad fleshy mass (4 mm. across) is thus formed which inserts on the base of the cranium between the bullae.

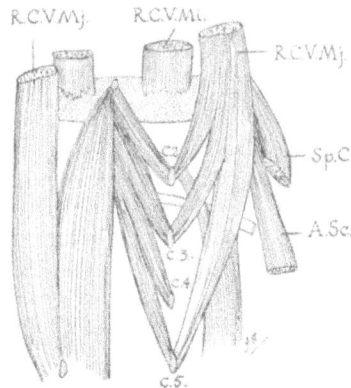

FIG. 19.   Dissection of prevertebral region of neck.

Rectus Capitis Anterior Minor

This is a small, short, quadrangular fleshy mass passing from the ventral aspect of the ventral arch of the atlas to the occipital bone, where it inserts posterior to the attachment of the preceding muscle.

Rectus Capitis Lateralis

A similar short, quadrangular fleshy sheet lying lateral and slightly posterior to the preceding, passing between the lateral mass of the atlas and the jugular process of the occipital bone.

Longus Colli

This consists of the usual three parts, as described by Woollard (1925) in *Tarsius* and Beattie (1927b) in *Hapale,* namely upper oblique, lower oblique, and vertical. The lower oblique fibres spring by tendinous fibres from the lateral aspect of the bodies of the anterior four thoracic vertebrae. Becoming fleshy, they incline laterad to insert on the ventral tubercles of the transverse elements of C.5 and 6.

The upper oblique fibres spring from ventral tubercles of the transverse elements of C.2–6, incline medially and insert into the median ventral tubercle of the atlas. Vertical fibres pass craniad from an origin common with lower oblique fibres to insert on the bodies of C.2–4 and a thin slip to the median ventral tubercle of the atlas. They are more fleshy than the lower oblique fibres enlarging to form a massive belly each side, occupying the concavity on either side of the median keel of the cervical centra.

9. SCALENE MUSCLES

Scalenus Ventralis

This arises from ventral tubercles of the transverse elements of the last three cervical vertebrae. The fibres converge to their insertion on the cranial surface of the first rib between the subclavian artery and the brachial plexus dorsally and the subclavian vein ventrally. There is no tendon, the muscle being fleshy to its insertion.

Scalenus Medius

This is not separable at its origin from the preceding but its fleshy fibres pass dorsal to the subclavian artery to insert on the first rib. The belly is pierced by the long thoracic nerve. Like the scalenus ventralis, it is fleshy throughout, but its fibres are slightly paler.

Scalenus Dorsalis (seu Posticus)

This is quite independent. It is an elongated, strap-like muscle, 34 mm. long by 2.5 mm. broad arising from the dorsal tubercle of the transverse process of C.4 only. It is fleshy all the way to its insertion, which is on the fourth rib. To gain its insertion the belly passes between the second and third digitations of serratus ventralis, by proceeding from the dorsal

(lateral) to the medial side of that muscle, and ventral to the intercosto-brachial nerve.

10. MUSCLES OF THE PHARYNX AND PALATE
(fig. 20)

The constrictor layer consists of three distinct entities, superior, middle, and inferior. These form three different strata overlapping from below upwards. The *superior constrictor* arises from the medial pterygoid plate and the side of the tongue in its posterior third and a few fibres from the mandible posterior to the last molar. The fibres sweep round the side walls of the pharynx on to its dorsal wall, meeting their fellows in an indistinct median raphe. The most cranial fibres are slightly oblique and approach the base of the cranium where the median raphe gains attachment to the basioccipital. The hindmost fibres are hidden by the middle constrictor and by the broad, ribbon-like insertion of the stylo-pharyngeus.

The *middle constrictor* clothes a shorter length of pharynx than the preceding or the inferior constrictor, but it is thicker, especially along its cranial border, which is slightly oblique, rising to a peak on the median dorsal raphe. Its origin is concentrated on the hyoid bone, the fibres spreading fanwise from the origin.

The *inferior constrictor* is not well defined below from the musculature of the oesophagus. Its origin is from a crest on the lateral surface of the ala of the thyroid cartilage, its superior cornu, and the thyro-

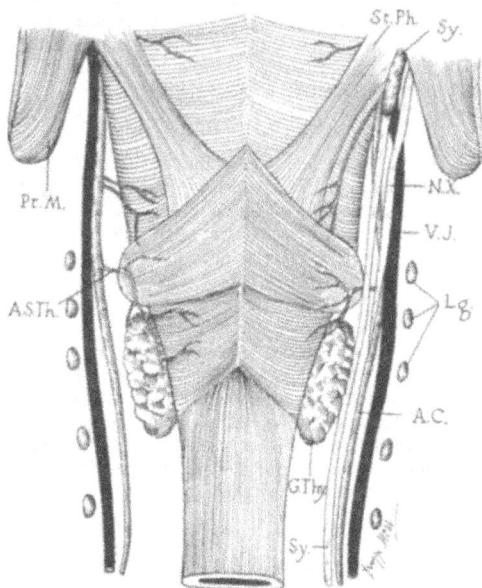

Fig. 20. Dissection of the pharynx from behind to show constrictor muscles, etc.

hyoid ligament. The upper fibres override the lower fibres of the middle constrictor but the lower ones blend over the lateral walls of the oesophagus with the longitudinal coat of that tube.

The constrictors show the usual relations to the laryngeal nerves and vessels.

The *stylo-pharyngeus* is a broad, ribbon-like sheet doubtless incorporating posteriorly some fibres from the palato-pharyngeus. It arises from the base of the tympanic bulla posteriorly and sweeps backwards and medially to enter the pharyngeal wall deep to the cranial border of the middle constrictor. Its fibres then spread out, some ending in the median raphe, others gaining the cranial border and superior cornu of the thyroid cartilage. It has the usual relation to the glosso-pharyngeal nerve, from which it receives its nerve supply.

The *palato-pharyngeus* arises from the posterior border of the hard palate and palatal aponeurosis. Its fibres converge forming a bundle contained within the posterior faucial pillar. Thereafter they spread out again and lose themselves in the pharyngeal wall, some becoming incorporated with the preceding muscle. A distinct salpingo-pharyngeus was not observed.

11. MUSCLES OF THE TRUNK

Supracostales are lacking. Intercostales call for no comment apart from their deep red color. The arrangements of the aponeurotic portions of these muscles are as in Man.

Transversus Thoracis

This covers the deep aspect of the sternum and xiphisternum as far anterior as the second rib. Fibres diverge to end in digitations inserting on the deep surfaces of ribs 2 to 7.

Rectus Abdominis

This springs by fleshy fibres from a line on the cranial edge of the pubis over a distance of 8 mm. It forms a dorso-ventrally flattened band, slightly narrower over most of its extent than at its origin. Passing forwards, it crosses the costal margin and proceeds on the thorax deep to the pectoral muscles as far forwards as the first rib. The belly shows no tendinous intersections. The lateral fibres insert on costal cartilages from the first to eighth. The medial part, which becomes more differentiated anteriorly, inserts on costal cartilages 2–5 and adjacent sternum. Its widest part is near the insertion on to the first rib, where it attains a width of 10 mm.

The rectus sheath is a firm fibrous envelope united to its fellow in the median line to form the linea alba. It is formed ventrally chiefly by aponeurotic fibres derived from the external oblique and dorsally by similar fibres from the internal oblique and transversalis, but also, over a narrow tract laterally, by continuation of fleshy fibres of the transversalis.

Obliquus Externus Abdominis

A thin sheet of pale fleshy fibres arising by eight digitations from ribs 5–12 (as in *Tamarin*), the anterior four interdigitating with the hindmost four slips of serratus ventralis. Posterior fibres of the last digitation are connected by strong fascia to the dorsal lamella of the lumbo-dorsal fascia and this fibrous connection is continued back all the way to the iliac crest; i.e. there is no free dorsal edge to the muscle. The most caudal part of the muscle gains a fleshy attachment to the anterior ventral spine of the ilium as in *Hapale,* but differing from *Leontocebus* (Windle, 1886) and *Tamarin* (Hill, 1957). Tendinous intersections (reported by Miller, 1947, in *Oedipomidas*) are lacking.

Between the iliac spine and the pubis there is no inguinal ligament, but the lateral half of the caudal edge of the muscle becomes aponeurotic, the aponeurosis being connected at intervals by fine though strong fascial fibres to the surface of the fascia lata over the base of the femoral triangle—as described for *Oedipomidas* by Miller. The medial half of the caudal edge is converted into a broad iridescent aponeurotic band forming an arch over the emerging spermatic cord. This arch, which has a short lateral and a much elongated and obliquely disposed medial column, is the "external abdominal ring." Its maximum length is 17 mm., and breadth 5.2 mm. The final fixation of the medial column or crus is to the pubis over the ventral surface of the origin of the rectus.

Anterior to the aponeurotic band just described the external oblique remains fleshy as far as the lateral border of the rectus; here it gives rise to a thin aponeurosis which proceeds to the linea alba, forming the ventral wall of the rectus sheath.

Obliquus Internus Abdominis

This is composed of fibres equally thin and pale as its superficial neighbor, except in the inguinal portion, where the fibres are ruddier in color. Fleshy fibres spring from (a) the ventral border of the ilium in its anterior half, (b) from the surface of the fascia lata over the lateral half of the base of the femoral triangle, and (c) from the lumbo-dorsal fascia along a line corresponding to the lateral border of the ilio-costalis muscle.

Most of the fibres proceed downwards and forwards towards the linea alba, while the anterior ones become attached to the costal margin where this is formed by the last five ribs. Between the eighth rib and the pubes the fleshy fibres end in aponeurosis at the lateral border of the rectus and over the greater part of this extent the aponeurosis divides, one layer passing deep and the other superficial to the rectus. From a point half-way between the umbilicus and the pubes, however, the whole of the internal oblique aponeurosis proceeds dorsal to the rectus.

Fleshy fibres of the most caudal part of the muscle

take a different direction, arching over the inguinal canal to end medially in a conjoint tendon with the most caudal fibres of the transversalis. The tendon inserts on the pubic bone immediately laterad of the lateral border of the rectus origin, and in close relationship to the medial surface of the spermatic cord.

Cremaster

This is a well-developed structure comprised of fleshy fibres proceeding over the ventral and lateral surfaces of the spermatic cord and derived from the adjacent caudal border of the internal oblique and transversalis muscles.

Transversalis Abdominis

This arises by digitations from the inner surfaces of the hinder six ribs, interdigitating with the diaphragm, from the lumbo-dorsal fascia in common with the internal oblique and from the iliac crest, anterior ventral iliac spine and lateral half of the crural fibrous arch.

The anterior fibres course almost directly ventrad or with a slight caudad trend, but the posterior ones arch somewhat forwards, those from the crural arch joining the conjoint tendon alongside the fibres on internal oblique. The muscle remains fleshy for a greater distance ventrally than the obliques, forming part of the dorsal wall of the rectus sheath, thereafter becoming aponeurotic. It gives a substantial contribution to the cremaster.

Quadratus Lumborum

This is an elongated muscle, much longer (57 mm.) than broad (12 mm.), occupying the interval between the last rib and the iliac crest. It is considerably wider anteriorly than at its iliac end, though the latter is commonly treated as the origin. Definitely laminated, the medial portion consists of three superimposed flattened bellies of isosceles-triangular outline, terminating posteriorly in long, glistening tendons, the largest being the most ventral, which terminates on the transverse process of the last lumbar vertebra and the others at successively more anterior levels (transverso-costal component). On a more dorsal plane is a broad fleshy lamina attached in front to the last rib and behind to the iliac crest. This part projects laterad of the transverso-costal fibres and is covered ventrally by the ventral lamina of the lumbo-dorsal fascia reinforced with parallel oblique aponeurotic bands passing from in front backwards and laterally in line with the emergent lumbar nerves.

12. DIAPHRAGM

This has the usual division into sternal, costal, and vertebral components. Sternal slips are thin, but broad and separated from each other and from the most ventral costal slip by clefts occupied merely by areolar tissue. Costal slips, six in number each side, spring from the inner surfaces of the last six ribs and interdigitate with slips of the transversalis. The crura are elongated fleshy bundles, ending posteriorly in slender glistening tendons, that on the left proceeding as far as the body of the fourth lumbar vertebra, while the right extends a little farther—to the disc between L.4 and L.5. There is a marked triangular lumbo-costal hiatus between the crural and costal parts each side, filled by fascia on the abdominal side derived from that covering the psoas.

The central tendon consists of a median ventral, more or less quadrate and imperforate portion with paired, narrow dorsal extensions. The median portion projects ventrad slightly more on the left than the right and its angles are rounded. It receives the insertion of the sternal slips and the more posterior costal slips. In the angle between the two dorsal extensions the fibres from the crura are received. The remaining costal slips insert on the lateral edges of the tendon. The total dorso-ventral extent of the tendon measures 31 mm., of which, on the left 21 mm. are accounted for by the dorsal extension. The ventral quadrate portion measures 12.5 mm. transversely. The caval opening is through the central tendon at the medial part of the base of the right dorsal extension. Dorsal to this and slightly to the left, the oesophagus perforates the fleshy part of the diaphragm in the interval between the two dorsal extensions. The aortic opening is still farther dorsally, between the two crura.

On either side of the central tendon, the fleshy portion projects as a dome into the thorax. The left cupola rises to the level of the fifth rib, and the right a shade beyond.

13. DEEPER MUSCLES OF THE BACK

The two *dorsal serrati* are almost entirely degenerate, being represented for the most part by glistening aponeuroses—especially the anterior one. Serratus dorsalis anterior arises from the spines of C.7 and the first two thoracic vertebrae, and inserts on the second to fifth ribs. The posterior serratus is composed of five aponeurotic slips passing from the lumbo-dorsal fascia to ribs 7 to 11, and a thin fleshy sheet behind this inserting along the hinder borders of the last two ribs.

Ilio-Costalis

A fleshy column divisible into lumbar and thoracic parts, the cervical portion being absent as in *Macaca* (Lineback, 1933: 121), though indicated in his figure 41. Ilio-costalis lumborum contrasts markedly with the neighboring longissimus column which lies medial to it, in that the latter is covered by strong aponeurotic fibres. Fleshy fibres arise from the iliac crest and from the intermuscular septum between it and longissimus. Caudal fibres pass somewhat obliquely laterad and forwards, encircling round the deeper fibres to insert on lumbar costal processes. The deeper and more cranial fibres are more sagittally disposed and insert largely on the last rib. Ilio-costalis dorsi is the forward continuation on the thorax. It is more flattened and tends to divide into slips which have tendinous ends. The general arrangement is for slips to arise from the hinder 6 ribs and to insert on the anterior 6, but there is some overlapping and interlacement.

Longissimus

This column is also fleshy, but is marked dorsally by a strong aponeurosis which contains locally further reinforcement by glistening tendinous bands or sheets having a generally forward and laterad trend over two or three segments. Over the depression between sacral spines and ilium the glistening appearance is continuous, but a separate tract passes from the spine of the last lumbar forwards and laterally to the anapophysis of the fourth. More anterior slips pass over a smaller number of segments; fleshy intervals occur between them, but in the thoracic region the aponeurosis again becomes continuous. The origin of the fleshy bundles is from the sides of the sacral and lumbar spines, from the dorsum of the sacrum and sacro-iliac joint and from the overlying fascia. Fibres pass cranially to insert on lumbar vertebrae, especially to the anapophyses and to the tendons of the deeper muscles (e.g. multifidus). There is a continuation in the dorsal region and also in the neck (longissimus capitis, already described, p. 42). The thoracic portion has additional insertions on the ribs near their tubercles.

14. MUSCLES OF THE TAIL (fig. 21, 51)

These consist of the following:

1. Extensor caudae medialis
2. Extensor caudae lateralis
3. Abductor caudae medialis
4. Abductor caudae lateralis
5. Intertransversarii caudae
6. Flexor caudae brevis ⎫
7. Flexor caudae longus ⎬ = sacro-coccygeus
8. Ischio-caudalis (= coccygeus)
9. Ilio-caudalis ⎫
10. Pubo-caudalis ⎬ = levator ani
11. Pubo-rectalis ⎭

The first five need no special description as they differ not at all from those of other tailed Primates. The remainder need some further consideration and it is convenient to take them in the reverse order as listed.

The muscles corresponding to *levator ani* of human anatomy form a very thin sheet of darkish fleshy fibres arising along a continuous line commencing on the pelvic aspect of the body of the pubis and continuing along the pubic ramus and beyond that on to the body of the ilium as far as the sacroiliac joint (as in all tailed monkeys; Kollmann, 1894). The most

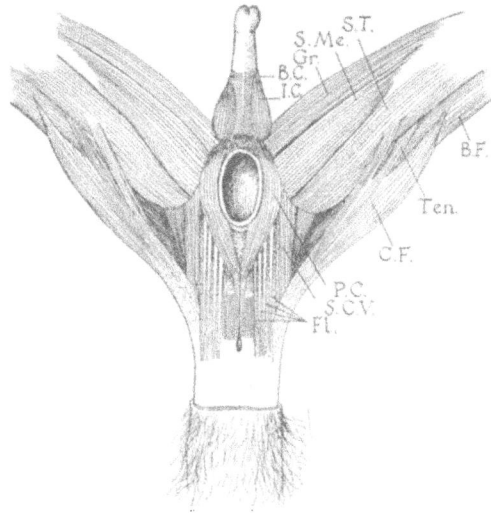

FIG. 21.　Dissection of the perineum to show muscles of tail, etc.

ventral fibres proceed caudad and slightly dorsad, embracing the lower urogenital tract and anal canal, superficial to the external sphincter, inserting on the wall of the anal canal very close to the anal margin. They are distinct from the next portion or pubo-caudalis, which arises from the horizontal ramus, constituting a pubo-rectalis—a muscle not present in *Tamarin* or *Hapale*.

These fibres have a more dorsal trend and join those of the ilio-caudalis just prior to the point where the latter become tendinous without gaining attachment thereto. Ilio-coccygeus is composed of more longitudinally directed fibres and is somewhat more bulky. Just posterior to the dorsal anal margin the fleshy belly suddenly becomes tendinous, receiving at the same time the fibres of pubo-caudalis. The tendons of the two sides approach and run distad for some five millimeters to insert on the remnant of the chevron bone of the third and fourth caudal vertebrae.

Ischio-coccygeus

This is not a dorsal representation of the levator ani sheet, but an independent muscle on a more dorsal plane. It is very small, triangular, and passes from the ischial spine to the transverse process of the first caudal vertebra.

Flexores Caudae Breves et Longi

These are not separable at their origin, which ascends as far craniad as the sacral promontory. The two form a fleshy mass in the sacral concavity, the muscles of the two sides arranged as a shallow trough in which lies the dorsal aspect of the rectum. Towards the apex of the sacrum the two become differentiated by a sagittal division, the short muscle being more medial. Its fleshy belly narrows to a tendon which inserts on the same tubercle as the ilio-caudalis a little more laterally. The longer muscle breaks up into four glistening tendons which proceed to the next four caudal vertebrae to that which receives the flexor brevis insertion.

15. MUSCLES OF THE PERINEUM (fig. 21)

Bulbo-cavernosus consists of a series of parallel fibres arranged in V-formation on the inferior surface of the root of the penis, with the apex of the V pointing caudad. It serves as a compressor of the posterior urethra.

Ischio-cavernosus is a thick fusiform fleshy mass on each side of the root of the penis, covering the corresponding crus, arising from the ischio-pubic ramus and inserting by tendon on the corpus cavernosum, laterally and dorsally. It clearly assists in erection of the organ. There is no distinct continuation comprising a levator penis muscle.

16. APPENDICULAR MUSCLES

A. Muscles of the Pectoral Girdle and Limb (figs. 22–32)

a. Dorsal Group

(1) Superficial layer (fig. 22). *Trapezius*—in contrast to that of *Hapale* and *Leontocebus* (Windle)—does not gain the occiput. Its highest point of origin is the spine of C.4. From here its origin extends continuously along the tips of the spinous processes and interspinous ligaments as far as the eighth thoracic spine. The fibres are mainly fleshy, but, as in *Tamarin*, a narrow fusiform speculum rhomboideum of aponeurotic fibres is located opposite the spines of C.5–T.1. The cervical fibres proceed obliquely ventrad and caudad to insert on the lateral half of the clavicle and acromion process. These are succeeded by less oblique and then by horizontal fibres passing to the anterior edge of the scapular spine. The posterior thoracic fibres have an obliquely forward course and their insertion is concentrated on the most medial part of the scapular spine, both on its expanded root and the caudal edge nearby. The fibres lie superficial to the anterior horizontal fibres of latissimus dorsi.

Latissimus dorsi: A broad flat triangular fleshy sheet arising from the spinous processes of the hinder five thoracic vertebrae, from the lumbar aponeurosis and feebly from the last three ribs. There is no fleshy connection with the lumbar spines or iliac crest, or any origin from the posterior angle of the scapula, but there is a fascial connection at the last-mentioned site. The cranial border is not disposed in the coronal plane, but somewhat obliquely, proceeding forwards and ventrad. Near its origin the muscle is extremely thin and pale, but the fibres converge as they pass over the

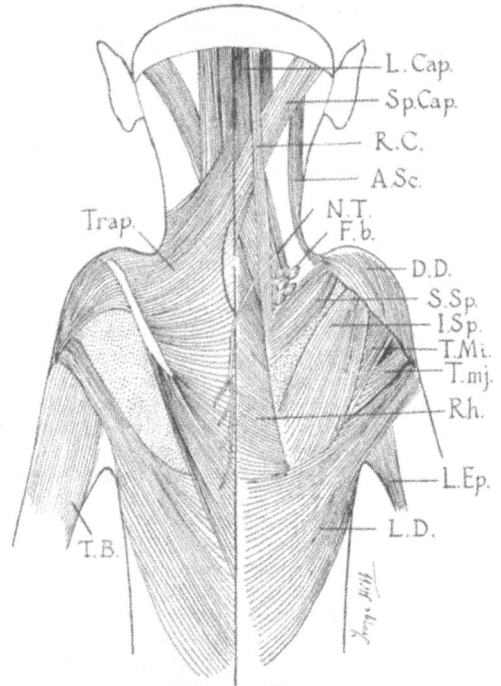

FIG. 22. Dissection of the back and shoulders showing on the left the superficial musculature and on the right the second layer of muscles.

posterior scapular angle to form a thick, flattened mass some 10 mm. across. This winds round the hinder border of the teres major, gives off the latissimo-epicondyloideus fibres as it does so, and inserts by tendon in the floor of the bicipital groove. Only the last 7 mm. is tendinous, forming a shining aponeurosis 4 mm. broad beneath the short head of biceps and coraco-brachialis.

(2) Deep layer (*Rhomboidei*) (fig. 22 right). *Rhomboideus capitis* is very distinct, though narrow and delicate. It is an elongated, pale, fleshy band, arising from the superior nuchal line of the occipital bone adjacent to the mid-line. The nuchal slip of *Hapale* is lacking. Proceeding sagittally along the neck, it thereafter diverges obliquely laterad and inserts at the antero-dorsal angle of the scapula superficial to the levator scapulae. It is crossed in its lower quarter by the nerve to trapezius. The muscle is 46 mm. long, but only 1.5 mm. wide.

Rhomboideus proprius is indivisible, as in most specimens of *Hapale*, *Tamarin*, and *Leontocebus*, though Beattie declared it to be divisible in *H. jacchus* into major and minor parts as in Man. The rhomboid sheet is 12 mm. broad and 20 mm. antero-posteriorly and its fibres are obliquely arranged. The origin extends from C.6 to T.4 and the insertion is concentrated chiefly near the posterior scapular angle, but some of the more cranial fibres proceed beyond the vertebral border on to the ventral aspect of the scapula, inserting on the fascia covering serratus ventralis.

Trachelo-acromialis: Another straplike fleshy tract lying deep to the ventral border of trapezius, between it and the deep aspect of the sterno-mastoid. It springs from the first two cervical vertebrae and inserts on the tip of the acromion, having no clavicular attachment.

b. Ventral Group (Pectorales) (fig. 23)

Pectoralis major lacks a clavicular origin. Its principal portion is arranged as in *Hapale*, the most anterior fibres transversely disposed and the posterior fibres obliquely, the latter rolling over the caudal border of the transverse portion to end in a flattened tendon on their dorsal aspect, but not crossing them. The origin is confined to the sternum, near the median line, and extends from the manubrium to the level of the seventh costal cartilage. The insertion (12.7 mm. broad) is mainly fleshy (except for the dorsally situated tendinous band which is applied to the hinder half of the fleshy portion) into the lateral lip of the bicipital groove on the humerus. The muscles of the two sides are separated by a narrow median fibrous tract.

A small slip is differentiated from the deep surface of the muscle as it arises from the manubrium. This forms a fusiform belly which inserts on the inferior aspect of the medial end of the clavicle and the fascia over the subclavius muscle. It evidently represents the sterno-clavicularis anterior, sometimes differentiated from the deep aspect of the muscle in Man (Bryce, 1923, in Quain, **4** (2): 102; also Huntington, 1904).

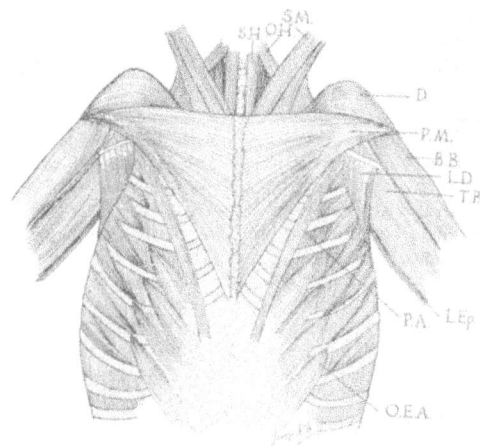

Fig. 23. Dissection of the pectoral region and root of neck.

Pectoralis abdominis is completely differentiated from the preceding, much more so than in *Hapale*, for there is a distinct interspace between them. Its origin is very thin and takes place over the aponeurosis covering the portion of the rectus sheath over ribs 8–9. A thin fleshy band is formed proceeding forwards and laterally, then deep to the pectoralis major to insert by a flat tendon to the capsule of the shoulder-joint over the head of the humerus, and on the fascia over the upper third of the long head of the biceps.

Pectoralis minor has a restricted sternal origin, deep to pectoralis major opposite the insertions of costal cartilages 3–6. It is fleshy throughout, the fibres converging to an insertion on the capsule of the shoulder joint.

Subclavius is a very robust muscle composed of darker fibres than the other pectorals. It arises by a short, stout tendon from the sternal end of the first rib. It inserts by fleshy attachment to the caudal surface of the clavicle over a distance of 11 mm.

c. Lateral Group

Levator scapulae: This and the next muscle form a continuous sheet, without any clear division, arising by a series of digitations from the dorsal tubercles of transverse processes of the hinder six cervical vertebrae and the anterior eight ribs. Levator may be defined as composed of the four most anterior digitations arising from the tubercles of cervical vertebrae 2 to 5, the slip from 2 being very slender and united immediately to the next. The digitations are tendinous at their origin, but rapidly become fleshy, losing their identity about half-way across the muscle. They insert mainly on the antero-dorsal angle of the muscle.

Serratus ventralis: The slips from C.6, C.7, and the first rib expand into a thin sheet which inserts along the vertebral border of the scapula in close relation to the insertion of rhomboideus (*q.v.*). The remaining five digitations are more bulky and converge to an insertion on the deep surface of the caudal angle of the scapula. The four posterior digitations interdigitate with the four most anterior slips of the external abdominal oblique. Between digitations 2 and 3, the scalenus dorsalis insinuates itself to gain its insertion on the fourth rib. The long thoracic nerve, which supplies the serratus, lies dorsad to the scalenus.

d. Deltoid Region

Deltoideus: A triangular muscle draped over the shoulder prominence, it consists of three distinct parts, cleido-deltoideus, acromio-deltoideus, and spino-deltoideus, each separated from the other by a fibrous septum, while a similar septum intervenes between the cleido-deltoideus and pectoralis major. The central portion, or acromio-deltoid, is a simple parallel-sided muscle, 6.5 mm. broad at its origin from the acromion process. Its fibres pass distad to end in a tendon attached to the deltoid impression on the lateral aspect of the humeral shaft. Fibres of the cleido-deltoid spring from the lateral half of the ventral surface of the clavicle and over a distance of 15 mm. proceed obliquely lateral and distad to insert on the septum along the ventral border of the acromio-deltoid. The spino-deltoid is larger (20 mm. along its base), extending dorsally along the posterior edge of the free margin of the scapular spine. Its fibres insert similar to those of cleido-deltoid along the fibrous septum on the dorsal border of the acromial portion, extending slightly less distad than those of cleido-deltoid. The insertion of the deltoid is embraced distally by the origin of the brachialis muscle.

e. Scapular Muscles (fig. 24)

Supraspinatus is a small fleshy mass occupying the supraspinous fossa. It is almost bipenniform, consisting of a group of straight fibres (pars propria) in its central part, receiving oblique fibres on either side, anteriorly from the cranial margin of the scapula, beyond which the fleshy mass bulges somewhat as a fusiform mass, and posteriorly from the cranial surface of the scapular spine. Insertion is on the highest point of the great tuberosity of the humerus.

Infraspinatus is a triangular flat sheet, very thin at its medial end, but thickening considerably towards the axilla. It is covered by a layer of dense aponeurosis from which some of its fibres take origin. It is incompletely differentiated from the teres minor, but well delaminated from teres major, which overlaps it to some extent posteriorly. Fibres continue to arise as far laterad as the neck of the scapula. All fibres converge to a stout tendon which inserts on the upper part of the posterior border of the great tuberosity of the humerus. Nerve supply is from the axillary nerve.

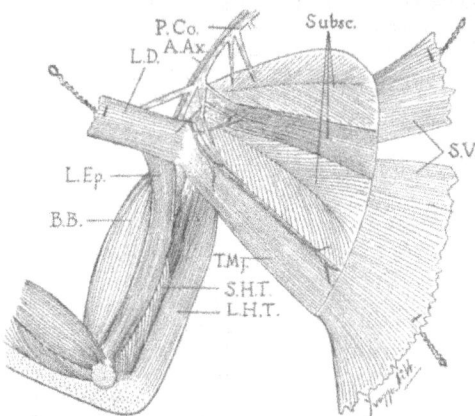

Fig. 24. Dissection of the scapular region and brachium from the ventral side after removal from the body.

Teres minor is a small fleshy mass arching across the inferior aspect of the shoulder joint. Its fibres spring from the axillary border of the scapula near the glenoid with a few from the infraspinatus aponeurosis. The tendon of insertion attaches to the most distal part of the great tuberosity. Nerve supply is from the axillary (circumflex) nerve.

Teres major is a large muscle of flattened form, becoming thicker towards its insertion. It springs from the axillary border of the scapula medial to the origin of teres minor and also from the infraspinatus fascia and from the surface of the posterior part of subscapularis. Its broad tendon inserts on the posterior lip of the bicipital groove in close relation to that of latissimus dorsi. It is supplied by the most distal of the subscapular nerves.

Subscapularis is a mixed fleshy and tendinous mass occupying the venter scapulae. It consists of three parts imperfectly differentiated from each other, but each with a separate branch from the posterior cord of the brachial plexus. The central portion, like that of supraspinatus, consists of relatively straight parallel fibres passing transversely from the middle of the scapula across the shoulder joint to the lesser tuberosity of the humerus. Anteriorly a bipennate portion has its fibres converging to a strong intramuscular tendon which converges to the previous tendon at the common insertion. Caudally is a still larger portion, similarly bipennate, with an intramuscular tendon and also with a superficial aponeurosis along its caudal border, where it is intimately related with the teres major.

f. Flexor Muscles of the Brachium (figs. 24, 25)

Biceps brachii has an intracapsular long head and a short head arising in common with coraco-brachialis from the tip of the coracoid. The long head is much the larger, the short head being much compressed and medially placed. The two join at the level of the middle of the shaft and proceed to a stout tendon which is wholly inserted on the tuberosity of the radius, i.e. without connection with the deep fascia at the elbow.

Coraco-brachialis is a long muscle arising in common with the short head of biceps. Its fibres insert mainly by fleshy attachment to the medial surface of the humeral shaft, chiefly at its middle (coraco-brachialis medius), but an elongated slip, partly fleshy, partly tendinous, proceeds along the brachium to insert on the supracondylar process, crossing the median nerve (coraco-brachialis longus). The muscle has the usual relation to the musculo-cutaneous nerve, which supplies it as well as biceps and brachialis. This muscle differs from that in any of the hapalid genera so far studied where it is usually small, e.g. in *Tamarin* and *Leontocebus*, consisting of little more than coraco-brachialis brevis. *Hapale*, however, shows the coraco-brachialis longus (Beattie).

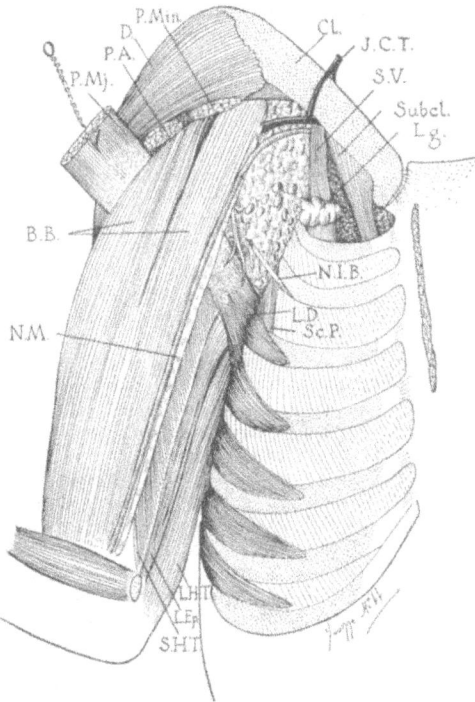

Fig. 25. Dissection of the right axillary and brachial regions.

Brachialis is, as in *Tamarin*, a transversely compressed fleshy mass, triangular in section with a sharp ventral border. Its origin ascends as high as the deltoid insertion, embracing it, especially on the lateral side. Its lateral aspect is applied to the brachioradialis. Its stout tendon of insertion gains attachment to the coronoid process of the ulna.

g. Extensor Muscles of the Brachium (fig. 25)

Triceps extensor brachii is similar to that of *Tarsius* as described by Burmeister (1846) and Woollard (1925) since the tendon of insertion on the olecranon receives fleshy fibres from 5 sources, namely the three normal heads, the latissimo-epicondyloideus (dorso-epitrochlearis) and the anconeus.

The most superficial contribution is that from the latissimo-epicondyloideus—a long ribbon-like fleshy band springing from the caudal border of the latissimus tendon and proceeding down the brachium at right angles to its parent. It fades out in the fascia over the medial aspect of the medial condyle and neighboring part of the triceps tendon and olecranon.

The long head of triceps is the bulkiest element. It springs by a very stout tendon 9 mm. across from the axillary aspect of the neck of the scapula and

adjacent part of the axillary border on a plane ventral to teres minor. It joins the lateral head at mid-shaft, the two proceeding thence to form the main part of the olecranal tendon.

The lateral head has the usual linear origin from the humeral shaft distal to the great tuberosity. It is partly fleshy, partly tendinous but soon becomes wholly fleshy.

The deep or medial head clothes the humeral shaft, from which it arises by fleshy fibres. It ascends almost to the surgical neck. It is narrow transversely and flattened dorso-ventrally. It contributes no fibres to the combined long and lateral heads until the final tendon is formed near the elbow.

h. Flexor Muscles of the Forearm and Hand (figs. 26–29)

Pronator teres has a single head of origin from the forepart of the medial epicondyle in common with the next muscle. An apparent deep head is a fibrous tract which is really part of the origin of flexor sublimis. The fibres constitute a fusiform fleshy belly which crosses the forearm obliquely to end in a flattened tendon inserting on a shaft of the radius about half-way down the forearm.

Flexor carpi radialis is a fusiform fleshy mass well differentiated from the preceding and almost as completely from palmaris longus. Its fleshy belly is succeeded, just distal to the level of insertion of pronator teres, by a straplike tendon. This becomes more terete as it passes the wrist, in a fibrous sheath, to insert on the base of the indicial metacarpal.

Palmaris longus is a fleshy mass lying postaxial to the preceding, arising from the more dorsal part of the epicondyle. It remains fleshy almost to the distal one-third of the forearm, being there replaced by a flat tendon which slowly fans out to gain attachments to the pisiform and the tubercle of the scaphoid while the middle portion proceeds more distad to fan out further on the palm, where it constitutes the palmar fascia.

Flexor digitorum sublimis is located on a slightly deeper plane than the preceding, being partly overlaid by it. It is also distinguished by its paler fibres. It consists of two portions, (1) a longer pre-axial part, not well differentiated from neighboring muscles at its origin from the epicondyle (deep to flexor carpi radialis and palmaris longus to which it is connected by tendinous slips) and (2) a very distinct postaxial belly of fusiform shape arising from the back part of the condyle deep to the palmaris longus. The flat preaxial belly divides on the lower third of the forearm, each part giving rise to a separate tendon. The two tendons proceed beneath the flexor retinaculum, the more radial one dividing again into two, contributing the perforated tendons to the second and third digits. The other tendon remains undivided and passes to the fourth digit. The second or fusiform belly con-

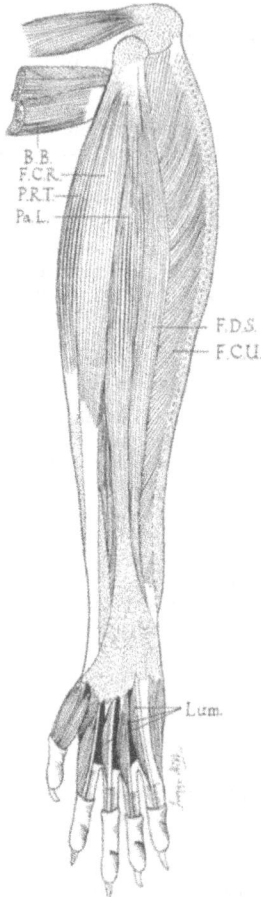

FIG. 26. Superficial dissection of the right forearm and hand.

stitutes an intermediary between flexor sublimis and flexor profundus. It gives rise to a stout tendon at mid-forearm, but this receives on its postaxial side, just above the wrist, an accession of fleshy fibres from the deep flexor. The tendon also gives fibres to the deep flexor tendons, only a small proportion of them continuing as the perforated tendon for the fifth digit.

Flexor carpi ulnaris is a broad flat sheet arising by two heads which span the interval between epicondyle and olecranon. From the olecranal head the attachment proceeds distad along the posterior border of the ulna whence fleshy fibres are picked up almost to the wrist. The muscle overlies the fleshy flexor profundus. Its tendon is short and stout, inserting on the large pisiform bone, whence the usual ligamentous continuations to the hook of the hamate and base of fifth metacarpal are formed.

Flexor pollicis longus: There is no deep flexor solely concerned with the pollex. Two thick fleshy masses cover the volar aspect of the shafts of the radius and ulna and intervening membrane. These are distinct from each other as far as their fleshy bellies are concerned, but there is a certain amount of interlacing of their emergent tendons, while the ulnar muscle also has connections with the flexor sublimis.

The radial component of the deep flexor mass, corresponding morphologically to flexor pollicis longus, has a multiple origin. Two distinct heads spring from the distal part of the medial epicondyle and are separated by the median nerve. In addition there is a fleshy origin from the volar aspect of the shaft of the radius and interosseous membrane. A thick tendon emerges in the lower part of the forearm, but this is inseparable medially from the correspondingly sized tendon of the flexor profundus proper. The radial tendon contributes slips to the pollex, index, and medius. It is supplied by the median nerve.

Flexor digitorum profundus: This is the ulnar moiety of the deep flexor mass. It consists of two main parts, superficial and deep, not fully separated from each other and having rather complicated interconnections between their tendons. The superficial or long head takes origin by fleshy fibres from the ulnar side of the orbicular ligament of the radius and from the volar surface of the proximal third of the ulnar shaft. It has no fibres from the interosseous membrane. In the middle of the forearm this head gives rise to a broad flat tendon which commences on the radial border of the fleshy belly, gradually widening to involve the whole muscle. The tendon becomes thicker towards the wrist. The fibres of the deep head are entirely fleshy and spring from the front of the ulna in the lower part of its shaft, fibres continuing to arise almost to the wrist. These proceed obliquely disto-preaxially and join the deep surface of the main tendon in bundles. The main tendon, as already noted, also receives a contribution from the flexor sublimis. In the carpal tunnel the flexor profundus tendon is intimately bound to that of the flexor pollicis, and like it becomes flattened and fan-shaped on gaining the palm. It finally divides into three supplying slips to digits III, IV and V. The middle digit therefore receives a contribution from both deep flexors.

In the palm the deep flexor tendons perforate the sublimis tendons in the usual way, finally inserting on the bases of the terminal phalanges. They are also related to the lumbrical muscles in the normal fashion.

Pronator quadratus: A thin lamina of dark fibres arising from the distal 8 mm. of the ulna. The insertion on the radius is slightly narrower.

i. Muscles of the Palm (figs. 26, 30)

As in *Hapale* opponentes are lacking from both pollex and minimus. The thenar eminence is formed by an abductor brevis pollicis and a flexor brevis, these being quite distinct though virtually continuous. The former springs from the trapezium and adjacent part of the flexor reticulum and inserts on the radial side of the base of the proximal phalanx of the thumb. Flexor pollicis brevis also arises chiefly from the trapezium slightly to the ulnar side of the preceding; it inserts on radial and ulnar sides of the base of the proximal phalanx. On its ulnar side lies the first contrahens muscle (= adductor pollicis).

On the hypothenar eminence it is difficult to determine the presence of more than one muscle. There is a single spindle-shaped fleshy mass passing from the

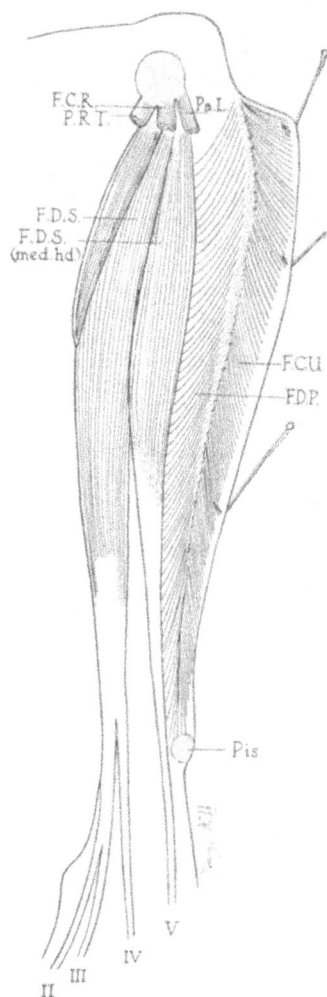

FIG. 28. Deeper dissection of the right forearm and hand specially to show the deep digital flexors.
FIG. 29. Deeper dissection of the right forearm and hand specially to show the flexor pollicis longus.

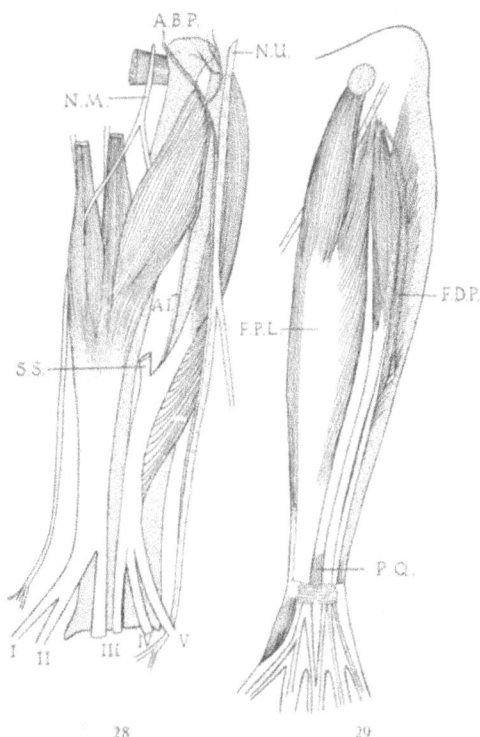

FIG. 27. Deeper dissection of the right forearm and hand.

hook of the hamatum and adjacent part of the flexor retinaculum to the ulnar side of the base of the proximal phalanx of the fifth digit. Near its origin this mass receives a bunch of short branches from the ulnar nerve as it turns towards the center of the palm.

In the central palmar compartment, besides the tendons of the long digital flexors, there are (1) the four lumbricales, (2) the contrahentes, and (3) the interossei.

Lumbrical muscles, four in number, spring from the radial sides of the deep flexor tendons supplying the ulnar four digits. They are slender fleshy bellies which proceed slightly preaxiad and distally, winding round the radial sides of the bases of the digits to insert on the dorsal tendon expansions. They are supplied by the ulnar nerve.

Contrahentes are also four in number, the first being the adductor pollicis. All arise from a fibrous expansion at the distal end of the carpal tunnel, deep to the long flexors. They comprise a very thin lamina of pale fibres which speedily diverge into four fusiform

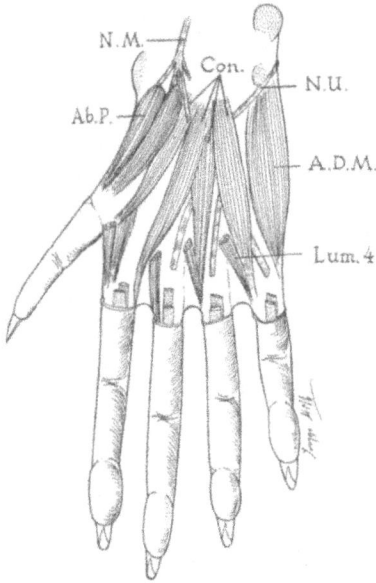

FIG. 30.   Deep dissection of the palm showing thenar and
hypothenar muscles and the contrahentes.

bellies, of which the most radial is more obliquely
disposed than the others. This belly is also more
distinct at its origin than the others. It inserts at the
ulnar side of the base of the proximal phalanx of the
thumb. The three ulnar contrahentes are subequal
in size. The middle one courses axially, but the two
flanking bellies radiate from it. Each inserts on the
radial side of the base of the corresponding digit. All
the contrahentes are supplied by the deep branch of
the ulnar nerve which crosses between them and the
underlying fascia which covers the interossei.

Callimico thus differs from *Hapale* and *Leontocebus*
where the contrahentes are absent, except for the
pollicial representative.

Interossei are divisible only with difficulty into palmar
and dorsal series, as in the Hapalidae.

j. Extensor Muscles of the Forearm and Hand
(figs. 31, 32)

Brachio-radialis and *extensor carpi radialis longior*
arise from the lateral supracondylar ridge of the
humerus, the latter over the distal 9 mm. of the ridge,
the former ascending to a point 16 mm. proximal to
the condyle. The remaining superficial extensors are
confined as regards their origin to the lateral epicondyle,
except the extensor carpi ulnaris, which bridges the
gap between epicondyle and olecranon. All are predominantly fleshy at their origin, but give place sooner
or later to tendon.

Brachio-radialis commences as a flattened fleshy
band, but distally it narrows, giving place to tendon
only in its terminal 3–4 mm. It inserts at the base
of the styloid process of the radius.

The two *radial extensors* are well differentiated
throughout, giving place to their respective tendons at
the junction of the middle and distal thirds of the forearm. The tendons are crossed superficially by the
strong tendons of the abductor pollicis longus. Thereafter the two tendons proceed in a single sheath over
the distal end of the radius to insert on the bases of
the second and third metacarpals respectively.

Extensor communis digitorum is a broad but thin
fleshy sheet which remains fleshy almost to the wrist.
Here it gives place to a broad tendon which broadens
still further on the dorsum of the hand, finally breaking
up into four main tendinous bundles distributed to the
four ulnar digits. There is, however, much criss-cross-
ing of fibres between these individual tendons, and also
with those of the next muscle.

Extensor digiti minimi is a fleshy band arising from
the lateral epicondyle dorsal to the preceding. It gives
rise to its tendon distal to the mid-point of the
antebrachium. This passes the wrist in a sheath of its
own in the groove between the radius and ulna. On
the dorsum of the hand the tendon bifurcates, the radial
division uniting with the extensor communis tendon
serving the fourth digit. The ulnar division passes to
the little finger, of which it is the sole extensor. This
arrangement is similar to that in *Hapale* and *Tamarin*.

Extensor carpi ulnaris: Arises by two heads, one
from the most dorsal part of the lateral epicondyle, the

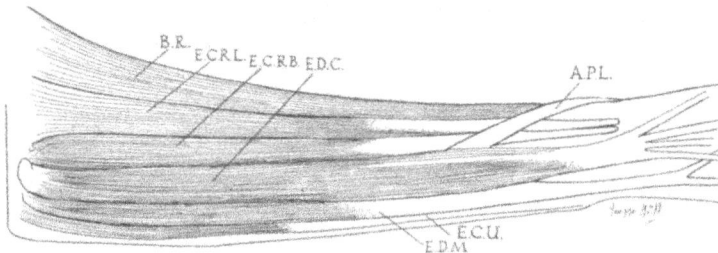

FIG. 31. Superficial dissection of the dorsal aspect of the forearm.

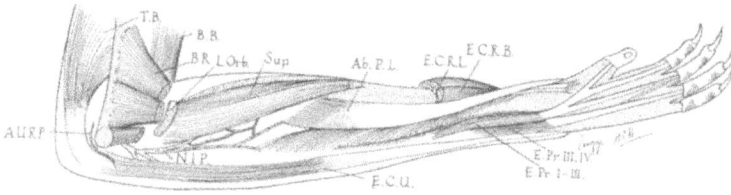

FIG. 32. Deep dissection of the radial side of the forearm and dorsum of the hand.

other from the lateral surface of the olecranon. No fibres are attached to the dorsal border of the ulna, but the muscle is covered by strong fascia which is so attached. The fascia does not give any attachment to fleshy fibres of this muscle. The fleshy belly is replaced by tendon from about the mid-point of the forearm. The final insertion is to the base of the fifth metacarpal. The tendon occupies its own compartment beneath the extensor retinaculum on the dorsal aspect of the head of the ulna.

Anconeus: Distinguished from the neighboring extensor muscles by its paler fibres, the anconeus consists of a few short fleshy strands connecting the lateral condyle and neighboring part of the supracondylar ridge with the border of the olecranon and edge of the ulna just distal thereto. Its origin extends over a distance of 8 mm. and its insertion about 5 mm. It is covered by dense aponeurosis continuous above with that over the triceps and below with the deep fascia of the forearm.

Deep muscles of the dorsal surface of the antebrachium include a supinator (brevis), abductor pollicis longus, and a deep extensor of the radial four digits. Extensor brevis pollicis is lacking as in the Hapalidae.

Supinator consists of relatively more longitudinal fibres than usual. These arise largely from the edge of the orbicular ligament of the radius, and only slightly from the lateral edge of the ulna, dorsal to the greater sigmoid notch. The fibres form a single fleshy stratum mixed with glistening aponeurotic fibres, proceeding distally and slightly preaxially to insert on the dorsal and medial surfaces of the shaft of the radius in its upper one-third. It is supplied by a twig from the radial nerve.

Abductor pollicis longus is a powerful muscle springing by two fleshy heads, one from the interosseous membrane near the insertion of supinator and the other, more elongated and unipenniform, from the dorsal surface of the shaft of the ulna and neighboring part of the interosseous membrane over the middle two-fourths of the bone. A broad, stout tendon commences very early in the more medial head, and to this the fibres of the ulnar head gain attachment on the deep surface, continuing along the ulnar border of the tendon almost to the wrist. The tendon narrows and becomes rounded, passes obliquely across the lower one-third of the radius, superficial to the extensores carpi radialis

longior et brevior, finally inserting on the base of the metacarpal of the thumb on its radial side. The muscle is supplied by the dorsal interosseous branch of the radial nerve, which does not pierce the supinator.

Extensores digitorum profundi: This term may be applied to the remaining two deep members of the dorsal group. They correspond to the extensor indicis of human anatomy, but have a more extensive origin and insertion. It closely resembles the corresponding muscle in *Tamarin*, being more differentiated than in *Hapale* (Hill, 1957).

Two bellies, arranged one distal to the other, arise from the dorsal aspect of the ulnar shaft near the posterior subcutaneous border, deep to the extensor carpi ulnaris and along a strip dorsal to the origin of the ulnar head of the abductor pollicis longus. Altogether the strip extends over the distal three-quarters of the ulna, and of this the distal one-third is accounted for by the smaller of the two bellies. Both bellies give rise to tendon just above the wrist. The tendon of the larger proximal belly lies to the radial side of the other. On the dorsum of the hand both tendons bifurcate, the contiguous divisions of the two tendons both proceeding to the medius, where they join the dorsal tendon expansion. The remaining (radial) division of the longer muscle as it approaches the heads of the metacarpals again divides, supplying contributions to pollex and index. There is thus no attempt at differentiation of a separate extensor pollicis longus, such as appears incipiently in *Tamarin*. The ulnar division of the tendon of the shorter deep extensor proceeds to the fourth digit.

B. Muscles of the Pelvic Girdle and Limb

a. Dorsal Group (Gluteal Region) (fig. 33)

The superficial layer comprises the tensor fasciae femoris, glutaeus maximus (ectoglutaeus), and caudo-femoralis.

Tensor fasciae femoris is much reduced, very thin and arises only from the region of the anterior ventral spine of the ilium. Its fibres sweep across the forward angle between flank and thigh, spreading somewhat as they go, and fading into the thick but transparent vagina femoris over the proximal part of the large vastus lateralis. The bipartite arrangement described by Jamieson (1904) in *Hapale* and reported also in

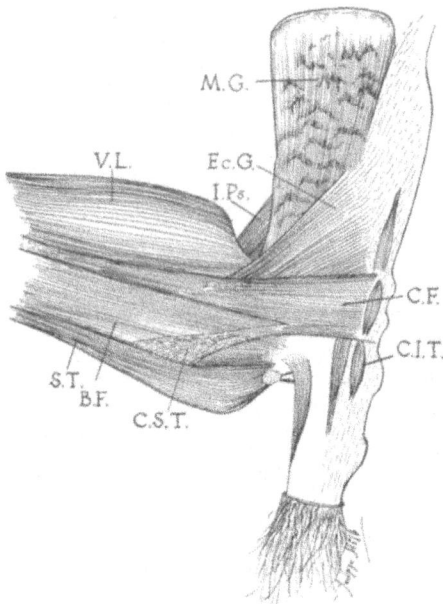

FIG. 33.   Dissection of the left gluteal region.

Tamarin (Hill, 1957) is not indicated and there is no connection, except through intervening fascia, with the ectoglutaeus.

As in other primitive Primates the glutaeus maximus is small and hence misnamed; the terms ectoglutaeus or glutaeus superficialis are to be preferred. It is a thin, triangular fleshy sheet of paler color than the other glutaei and the caudo-femoralis. Its base, formed by its origin, is from lumbar fascia and its backward continuation over the sacral region, opposite the hinder third of the ilium, to the dorsal border of which a few fibres are attached. They converge to a tendon which passes over the great trochanter and the lateral surface of the origin of vastus lateralis to insert along the site of the linea aspera over about 4.5 mm. of the femoral shaft in line with the caudo-femoralis, which passes more distad. The cephalic head described in *Hapale* is not present, but Windle (1886) has shown that there is some individual variation in the development of this muscle.

Caudo-femoralis is a well-developed structure with a more concentrated and thicker origin than the preceding. Its origin is in line with that of the ectoglutaeus but separated from it by a slight space. It has no origin deep to the ectoglutaeus such as occurs in *Hapale*. Its origin is by fleshy fibres from the third or fourth postsacral vertebrae. They proceed distally as a broad fleshy band which, beyond the trochanter, twists on its axis, the lateral (superficial) surface becoming pos-

terior and the deep surface anterior. Fibres insert consecutively along the posterior border of the femur in alignment with the insertion of the ectoglutaeus (see also below).

From the posterior border of caudo-femoralis are given off two fleshy slips, an anterior and somewhat deeper band and a posterior quite superficial one. The latter is the shorter and passes obliquely backwards and distally to join the biceps and semitendinosus, chiefly the latter. This evidently corresponds to the "vertebral head" of the latter muscle of *Hapale* (Jamieson, 1904).

The longer, deeper band proceeds medially and deeply between the biceps and the caudo-femoralis, from . which it is afterwards separated by the great sciatic nerve, from the peroneal portion of which it gains its own nerve supply. It, therefore, seems to be an aberrant tenuissimus, aberrant in that it inserts on the back of the caudo-femoralis instead of passing on to the fibula (as in *Tupaia*, Appleton, 1928; Howell, 1938; or *Cebus*, Green, 1931, or as found in *Tamarin*, Hill, 1957).

Glutaeus medius (mesoglutaeus) is a thick, powerful muscle lying beneath the gluteal aponeurosis from the deep surface of which many of its fibres take origin. The aponeurosis is attached all round the periphery of the gluteal fossa of the ilium and is continuous dorsally with the lumbo-dorsal fascia. The muscle is not attached to the whole gluteal fossa, its deep origin being confined to the anterior half of the dorsal border of the ilium and from the lumbar fascia covering the great sciatic notch, also from the dorsal ligaments of the sacro-iliac joint. The belly is almost completely fleshy, except for a slight tendinous formation superficially along its ventral margin; and it remains so to its insertion ventrally, but is replaced just prior to the insertion by tendon dorsally, the fleshy fibres curving over to insert on the tendon, the latter gaining its final attachment to the dorsal border of the greater trochanter.

Glutaeus minimus (endoglutaeus) is smaller than the preceding (*cf. Leontocebus*, Windle, 1886) but like it predominantly fleshy. Straplike in form, it arises from the rest of the very concave gluteal fossa of the dorsum ilii. It becomes tendinous only on approaching its insertion which is to the anterior (cranial) border of the trochanter major anterior to the pyriformis tendon. There is no scansorius (glutaeus quartus) or further subdivision of the endoglutaeus such as Jamieson describes for *Hapale*.

Pyriformis is large and fleshy, as Jamieson found in *Hapale*, arising from the pelvic aspect of the bodies of the three true sacral vertebrae. Fibres converge to form a rounded belly which narrows distally to a stout tendon inserting on the apex of the great trochanter. Its distinctness is in agreement with findings in Hapalidae and therefore contrasts with the state in most monkeys where it is fused to some degree as a rule with the glutaeus medius.

Obturator internus: This is not distinctly separable from the gemelli. It is a large fan-shaped fleshy sheet arising all over the pelvic aspect of the pubis and ischium and obturator membrane, except anteriorly, where the obturator nerve and vessels escape from the pelvis. The gemelli arise from the dorsal border of the ischium, one anterior to the tendon of the obturator as it turns over that border, and the other posterior thereto. The conjoined mass converges to its insertion by tendon into the medial aspect of the great trochanter. Tendinous fibres appear on the superficial surface of the central portion of the compound muscle. Gemellus inferior is somewhat larger than gemellus superior. The usual synovial bursa occurs beneath the tendon as it crosses the ischial margin. The muscle is crossed by the great sciatic nerve near the trochanter and by the nerve to the hamstrings more dorsally.

Quadratus femoris is a large muscle shaped like a parallelogram with long proximal and distal but shorter medial and lateral edges. Its origin from the ischial tuber and neighboring part of the ramus is located somewhat higher than its insertion, which is along a linear attachment of some 15 mm. distal to the root of the great trochanter on the lateral aspect of the femoral shaft. The muscle is wholly fleshy, all the fibres coursing parallel to the upper and lower borders. The distal fibres are quite distinct from those of adductor magnus, a vessel (medial femoral circumflex) passing between the two (cf. *Tamarin,* Hill, 1957).

b. Hamstrings (fig. 34)

These comprise the biceps femoris, semitendinosus, semimembranosus, and presemimembranosus. The general arrangement is the primitive mammalian one (Appleton, 1928) as in *Hapale.*

Biceps and *semitendinosus* arise by a common musculo-tendinous head from the lateral surface of the ischial tuberosity. The former has no second head. It broadens in the thigh to a flat fleshy band which becomes still broader at its insertion. This is partly fleshy over the lateral aspect of the capsule of the knee joint and head of fibula and more distally aponeurotic— into the deep fascia of the leg. The total breadth at the insertion is 28 mm., of which the fleshy portion accounts for 16 mm. Jamieson believed the fibular attachment to be peculiar to *Hapale* in Primates lower than the anthropoid apes.

Semimembranosus is 6.5 mm. wide just beyond its than at its insertion, which is by a long (12 mm.) rounded tendon which attaches to the upper end of the medial aspect of the tibial shaft deep to sartorius and gracilis. The additional head has already been referred to in dealing with caudo-femoralis. This head crosses the superficial surface of the biceps to reach semitendinosus.

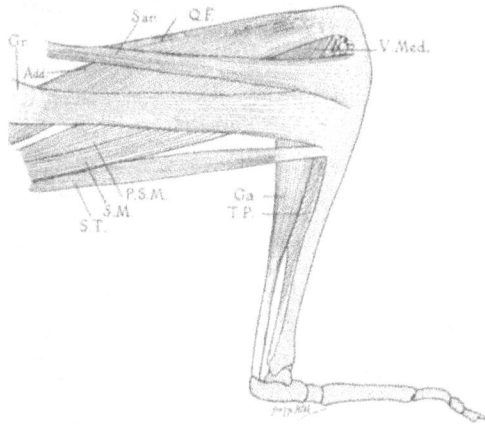

FIG. 34. Dissection of the medial aspect of the left thigh.

Semimenbranosus is 6.5 mm. wide just beyond its contracted origin from the ischial tuberosity medial to the semitendinosus and biceps. It expands to a fairly massive fleshy belly, but the dorsal three-quarters near the origin is membranous on the superficial surface. The muscle remains fleshy almost to its insertion where it suddenly gives rise to a short, but very stout, tendon, which gains attachment to the medial surface of the head of the tibia beneath the upper slip of the sartorius.

Presemimembranosus is quite distinct from the preceding and also, except for some tendency to fusion distally, from the adductor magnus. Its origin, flatter than that of semimembranosus, is placed more ventrally and is from the edge of the ischial ramus between semimembranosus and adductor gracilis, but a slight gap intervenes between it and the last-mentioned muscle, permitting the adductor magnus to be seen. It courses parallel to the semimembranosus, but its belly is triangular in section and crossed by the vessels and nerve proceeding to the gracilis. Insertion is into the adductor tubercle.

c. Adductor Group (fig. 34)

These are well differentiated compared with the reports given of their condition in other hapalids (Appleton, Jamieson, Windle, Beattie). They comprise a well-marked adductor longus, a gracilis, adductor brevis, adductor magnus, obturator externus, and a double pectineus. Of these adductor brevis is by far the bulkiest.

Adductor magnus is a broad flat, but thin fleshy stratum lying immediately anterior to the presemimembranosus and arising from the adjacent part of the ramus of the ischium. At its widest part it is 8 mm. across. At its origin a few glistening aponeurotic fibres affect the dorsal moiety, the remainder being fleshy. The fleshy insertion is marked by some divergence of the fibres which radiate in fasciculi to become attached over a fair area of the popliteal aspect of the medial condyle and neighboring part of the femoral shaft on a more dorsal plane than the other adductors and more proximal than the insertion of presemimembranosus.

Adductor brevis is a thick fleshy belly of pyramidal form arising from the pubic part of the ischio-pubic ramus by fleshy fibres throughout. The belly is 9 mm. broad and does not narrow appreciably towards the insertion, which spreads out in a linear attachment over some 24 mm. of the back of the femoral shaft in its middle.

Adductor longus is a flatter slenderer muscle some 5.2 mm. broad at its origin, but narrowing distally. The fleshy part is 48 mm. long. The insertion extends along the femoral shaft proximad and medial to that of adductor magnus over a relatively confined area.

Pectineus is formed by a thin fleshy lamina occupying a very oblique plane such that the femoral triangle (of Scarpa) is converted into a deep, narrow trough, the anterior surface of the muscle facing more laterad than in the human thigh. The muscle is distinctly divisible by a longitudinal cleft, occupied by areolar tissue, into a medial and a lateral portion. The gap widens distally, where the two portions diverge somewhat. The origin is narrow (5.2 mm.) and from the anterior border of the pubis. The medial portion is shorter than the lateral, whose fibres insert more distally on the femur. The total linear attachment at the insertion accounts for 17.2 mm. of shaft medial to the insertions of the other adductors.

Gracilis is the most superficial of the adductors. Very broad and partly tendinous at its origin, the muscle forms a broad but thin fleshy stratum over the medial surface of the thigh. The origin is along the medial edge of the ischio-pubic ramus superficial to the origin of adductor brevis and adductor longus over a distance of 12 mm. The belly inclines posteriorly as it courses down the thigh, narrowing slightly in the middle, but broadening again towards its insertion on the forepart of the medial surface of the upper end of the tibial shaft in a linear attachment 15 mm. long, superficial to the insertion of semitendinosus, but partly overlapped above by that of sartorius.

Obturator externus: The deepest of the adductors, this muscle is no less extensive than the obturator internus, arising from the obturator membrane and all margins of the foramen, including the pubic. The pubic slip is somewhat separate from the rest, with the obturator nerve passing between it and the remainder of the muscle. Fibres converge to a tendon which has the usual relation to the capsule of the hip joint, inserting in the floor of the digital fossa, hidden by the tendon of the obturator internus.

d. Coxal Flexors

Here are included the psoas and iliacus muscles; the sartorius and the quadriceps femoris complex.

Psoas parvus is well differentiated and accounts for most of the fibres arising from the lumbar vertebrae. Its fleshy fibres end in a broad flat tendon 34 mm. long, which narrows posteriorly to insert on the anterior border of the os innominatum medial to the fibres passing to the ilio-psoas insertion. It is crossed obliquely by the obturator nerve.

Ilio-psoas: Most of the fibres contributing to the short stout tendon which inserts on the lesser femoral trochanter are of iliac origin. The iliacus consists of a robust fleshy mass springing from the pelvic surface of the ilium. Most of the fibres end in the common tendon, but a few insert directly on the femoral shaft just distal to the trochanter. The lumbar contribution consists merely of a few fleshy strands from the hinder lumbar vertebrae, deep to the psoas parvus. There is certainly nothing resembling the description and figure given by Beattie for *Hapale.* Jamieson on the contrary agrees that psoas magnus is relatively small in *Hapale,* though connected with lumbar vertebrae 3–6. I have not confirmed for *Callimico* the curious relation between iliacus and quadratus lumborum met with by Jamieson in *Hapale;* the slip on the medial side of the quadratus is formed by the slender psoas contribution to the main muscle. There seems to be no iliacus minor or ilio-capsularis.

Sartorius is a very attenuated pale band some 10 mm. broad. Its origin is from the posterior half only of the ventral border of the ilium, i.e. it fails to gain the anterior ventral spine, but it receives additional fibres from the lateral part of the crural arch. Fibres run parallel to each other, coursing obliquely and superficially over the medial aspect of the thigh. At the knee they tend to spread out and become membranous, two fleshy divisions being defined, an upper which passes to the uppermost part of the head of the tibia, and a larger principal insertion to the medial aspect of the upper end of the tibial shaft, slightly overlaying the uppermost part only of the gracilis insertion (fig. 34). This agrees more with the condition in *Leontocebus* than in *Hapale.* The origin, on the other hand, recalls the condition in lemurs rather than that in Hapalidae.

Quadriceps femoris: The bulk of this muscle is formed by the huge vastus lateralis and the moderately developed rectus femoris. Vastus medialis and intermedius are feeble.

Rectus femoris has the dual origin by straight and reflected heads as in Man, but the two are not far apart. The muscle expands to a fusiform, but flattened belly, in which the fibres are mainly parallel. An aponeurosis appears on the deep surface towards the distal end. Finally the whole muscle becomes a broad flattened tendon which inserts on the proximal border of the patella.

Vastus lateralis recalls that of *Tarsius* though not hypertrophied to the same degree as in that genus. Commencing from a strong fibro-muscular attachment to the lateral aspect of the greater trochanter, the muscle swells up abruptly to form a medio-laterally flattened but antero-posteriorly thickened belly which is molded upon the lateral surface of the femoral shaft without gaining any direct attachment thereto. Distally it ends in a tendinous expansion which inserts on the lateral part of the proximal border of the patella and the whole of its lateral border. Many fibres pass superficial to the patella to join in the ligamentum patellae.

Vastus medialis is much smaller, arising from the front of the femur adjacent to but distad of the lesser trochanter, and the region in front thereof (anterior intertrochanteric line). It has few tenuous connections with the medial intermuscular septum. Its fibres course obliquely medially and distally to end in a fibrous expansion continuous laterally with the rectus tendon inserting alongside it and on the upper part of the medial border of the patella.

None of the three muscles just described has any connection with the accessory cartilaginous nodule which lies proximal to the patella over the trochlear surface of the femur. This body has been noted by Jamieson in *Hapale* and by Retterer and Vallois (1912) in *Tarsius* and *Leontocebus*, and by Hill (1957) in *Tamarin*. Retterer and Vallois describe the structure as an accessory patella, noting its variability of constitution (vesiculo-fibrous, cartilaginous, or ossified). In *Callimico* it is scarcely beyond the first-mentioned category and is therefore less developed, though extensive, than in the Hapalidae, being certainly less evident than in *Tamarin midas* where it forms a prominent structure. It is developed solely in a stratum of fibrous tissue lying wholly deep to the three structures described above as inserting on the patella proper, being developed in the fibrous expansion of the vastus intermedius muscle—as found by Retterer and Vallois in their examples.

The intermediate vastus takes origin from an elongated triangular area on the front of the femoral shaft in its distal two-thirds. The apex of the triangle is located proximally. The muscle ends below in the fibrous expansion mentioned in the preceding paragraph. The muscle receives several branches from the femoral nerve. No subcrureus (articularis genu) is differentiated.

e. Muscles of the Calf (figs. 35, 36)

Gastrocnemius is not particularly powerful. It has the usual bicipital origin, each head with a sesamoid bone (fabella) applied to the top of the corresponding femoral condyle and provided with an articular facet for gliding thereon. The two heads are of about equal

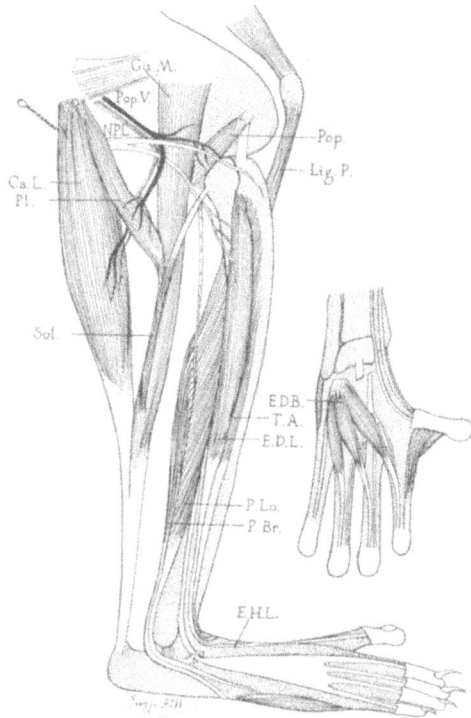

Fig. 35. Dissection of the calf muscles and extensors of the right leg, and dorsum of foot (inset shows deeper dissections of dorsum of foot).

size and both bellies are strongly compressed mediolaterally in their upper portions. Below they become broadened to some extent, more especially the lateral one. They remain separate and fleshy as far as the middle of the crural segment, thereafter being replaced by the powerful elongated, antero-posteriorly compressed tendo Achillis, which inserts on the tuber calcanei.

Plantaris arises in common with the lateral head of gastrocnemius, somewhat more deeply. Its fusiform fleshy belly passes obliquely medially and fuses with the lateral aspect of the medial belly of the gastrocnemius some few millimeters prior to the fusion of the two gastrocnemius bellies. The muscle does not give rise to a tendon and has no direct connection with the tendo Achillis as is usual in Hapalidae.

Soleus is a feeble structure attached by a long, narrow, flattened tendon to the back of the head of the fibula. It expands to form a fusiform fleshy belly which gradually unites with the medial belly of gastrocnemius after the latter has received the plantaris.

Popliteus presents no unusual feature, arising by

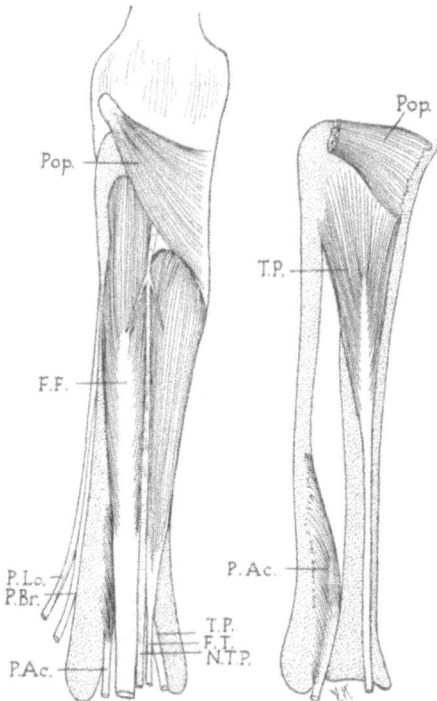

FIG. 36. Two deeper dissections of the calf region from behind.

tendon from the lateral surface of the lateral femoral condyle, passing beneath the lateral collateral ligament of the knee, then expanding into a triangular fleshy sheet which fans out over the upper part of the posterior aspect of the tibial shaft proximal to the oblique line.

Distal to the popliteus and on the same general plane lie the two long digital flexor muscles, flexor tibialis and flexor fibularis, of which the latter is distinctly the larger, being both longer and bulkier. The groove between their respective bellies is occupied by the posterior tibial nerve, which supplies branches to both muscles.

Flexor fibularis arises by fleshy fibres from the dorsal aspect of the fibula from an area distal to the origin of soleus, and for some indecisive distance distad thereto, but the distal attachment, apart from connections with related intermuscular septa, is so feeble as to be readily stripped off during dissection. The posterior surface of the muscle presents a broad glistening tendon in its distal two-thirds. This commences in the middle of the breadth of the fleshy belly, but gradually increases until the whole muscle is replaced by a single stout tendon, oval in section. This passes through the flexor retinaculum at the ankle to enter the

sole of the foot. At the ankle it lies lateral to the flexor tibialis, but in the sole it lies deep to it. The two tendons become flattened and broadened and fuse in the middle of the sole opposite the heads of the meta-tarsal bones. Flexor fibularis supplies tendons to all the digits except the fifth. Before the terminal tendons emerge the conjoint parent tendon forms a broad triangular sheet over the metatarsal region.

Flexor tibialis is shorter than the preceding by virtue of the greater distal extension of the popliteus on the tibial side of the leg. The muscle arises from the oblique line and from an uncertain distance along the posterior surface of the tibial shaft near its medial border. The fleshy belly becomes tendinous a few millimeters proximal to the ankle, the tendon being smaller and rounder than that of flexor fibularis. In the sole it lies superficial to the flexor fibularis and it supplies the tendon to the fifth digit, with a small contribution to that destined for the fourth digit. The independence of the tendon to digit V recalls the condition recorded by Windle in *Leontocebus*.

A third layer of muscles in the calf comprises a moderate-sized tibialis posterior and a small fusiform belly clothing the distal fourth of the fibular shaft. The latter terminates below in a slender tendon which proceeds to the hallux. It appears to be differentiated from the flexor fibularis and is so developed on the left side only.

Tibialis posterior is larger proportionally than in *Hapale* or *Tamarin*. Arising principally from the back of the tibial shaft below the oblique line and lateral to the origin of flexor tibialis, it also extends across the interosseous membrane on to the medial part of the upper quarter of the fibular shaft. Its fleshy fibres converge to a tendon which commences in the center of the fleshy belly, emerging from the muscle somewhat above the mid-point between knee and ankle. The long thin tendon appears superficially to the tibial side of the flexor tibialis tendon, entering the sole of the foot by passing beneath the flexor retinaculum in close apposition with the medial malleolus. It grooves the talus and proceeds in the sole deeply to insert mainly on the navicular, but with extensions to the two lateral cuneiforms. The arrangement agrees with that in *Ateles* as described by Förster (1922a).

f. Muscles of the Sole (figs. 37, 39)

Flexor digitorum brevis: A small muscle with a single head of origin and supplying tendons distally to two digits only—II and III. The origin is from the tubercle of the calcaneus, with some attachment to the plantar fascia superficially and to the surface of the long flexor tendon deeply. The belly is flattened proximally, bifurcating distally into two fusiform bellies which narrow rapidly to thin, rounded tendons which proceed distally with the long flexor contributions to II and III, the latter eventually perforating the flexor brevis tendons in the usual way. Possibly the loss of bellies to IV and V is attributable to damage done in removing the skin, but there is known to be considerable individual variation among Primates in the development of this muscle (Straus, 1930) albeit the variations relate rather to the relative development of the deep and superficial heads and to their respective contributions to the tendons supplying the four postaxial digits. Furthermore it may be recalled that Windle (1885–1886) recorded but a single tendon (supplying the index) in *Leontocebus*.

Musculus accessorius (quadratus plantae): This is a well-developed fleshy sheet of rather paler fibres than the other deep muscles of the sole. It springs from both tubercles at the plantar end of the tuber calcanei and the fleshy fibres proceed in an oblique direction towards the center of the sole. They insinuate themselves between the tendons of flexor tibialis and flexor fibularis, becoming finally attached to the former. It was not possible to trace any of its fibres directly to the hallux in view of the separate nature of the hallucial flexor on the left and its insignificance on the right, where it diverged from the flexor fibularis mass at right angles to the main course of the fibres supplying other digits.

Lumbricales: Only three are present, that for the index being absent. They are flattened fusiform bellies arising from the angles between the diverging long flexor tendons supplying the four post-axial digits, and also, to some extent, from the plantar surface of the main tendon sheet where they form a continuous fleshy stratum. Their slender inserting tendons proceed to the preaxial sides of digits III, IV, and V and attach themselves to the dorsal tendon expansions. They are supplied by the lateral plantar nerve.

Short muscles of the hallux: These are feebly developed though not absent as Beattie declares to be the case in *Hapale*, with the exception of a small abductor. In *Leontocebus* Windle found an abductor, a flexor brevis and a weak adductor.

Abductor hallucis is feebly developed. It arises from the plantar aponeurosis only, having no true bony origin, a condition recorded elsewhere in *Lemur* (Cunningham, 1882; Murie & Mivart, 1872) and *Daubentonia* (Zuckerkandl, 1898). It inserts at the preaxial side of the base of the proximal phalanx, but a sesamoid intervenes.

Flexor brevis hallucis: This is a single slender fusiform belly arising from the medial tubercle of the tuber calcanei. It proceeds obliquely distad across the sole, superficial to the long flexor tendons and ends opposite the origin of the preceding muscle in a long slender tendon which proceeds to the base of the proximal hallucial phalanx.

The presence of only one (the tibial) head again recalls the condition in *Lemur* (Cunningham, 1882). Usually two heads occur in both New and Old World Monkeys (Kohlbrugge, 1897; Ruge, 1878; Bischoff, 1870).

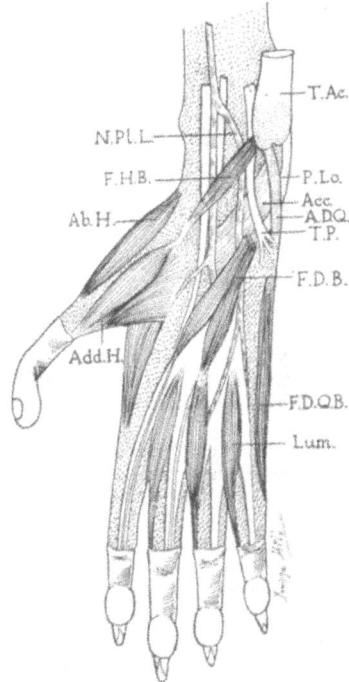

FIG. 37. Superficial dissection of the right sole.

Functionally the muscle is an opponens; a true opponens is not morphologically distinguishable.

Adductor hallucis: This is a very attenuated, though distinct muscular, slip arising from the sheath of the peronaeus longus not far from the latter's insertion. It passes obliquely to the hallux, inserting on the fibular aspect of the base of the first phalanx. Superficially it appears, at first sight, to be a distal continuation of the peronaeus, but careful scrutiny shows its distinctness. It is differentiated from the contrahentes layer, but quite separate from the other members of the group. There is no transverse adductor.

Contrahentes pedis: To the lateral side of the preceding is a very thin, almost membranous layer of muscular fibres of triangular outline lying between the long flexor tendons superficially and the fascia covering the interossei deeply. The origin is pointed and finds attachment to the sheath of the peronaeus longus and heads of metatarsals II–IV. Distally the sheet expands and gives rise to slips which proceed to digits II and IV only. The muscle is supplied by the lateral plantar nerve. The loss of the contrahens to digit V is unusual and negatives the phylogenetic order of disappearance postulated by Straus (1930).

Short muscles of the fifth digit (figs. 37, 39): A thick fibrous strand passing from the lateral tubercle of the calcaneus to the styloid process of the fifth metatarsal represents Wood's muscle (abductor ossis metatarsi quinti digiti).

A flexor brevis is present as a pale fleshy slip with the usual attachments. It lies in apposition to the most lateral of the interossei.

Abductor digiti minimi is a long slender fleshy band, composed of strictly parallel fibres, lying along the lateral border of the foot. It commences where the above-mentioned fibrous vestige of Wood's muscle terminates and proceeds to the metatarso-phalangeal joint, where it is attached to the sesamoid on the fibular side of the joint-capsule.

Pedal interossei (fig. 39) : Two muscles occur in the first interdigital space, a dorsal, with two heads, and a plantar. They both lie against the tibial side of the shaft of the metatarsal of the second toe and insert on the medial sesamoid opposite the metatarso-phalangeal joint.

Five other interossei are differentiated, but it is difficult to say whether they represent dorsal or plantar series. They all lie in much the same plane, which is decidedly plantar in position. One pair lie opposite each other along the shaft of the third metatarsal, inserting one on each of the two sesamoids in the capsule of the metatarso-phalangeal joint of that digit. A second pair, similarly arranged, is associated with the fourth digit. The remaining muscle is a slender one located along the tibial side of the shaft of the fifth metatarsal and inserting on the tibial sesamoid of that digit. All the interossei are supplied by twigs for the lateral plantar nerve.

g. Peronaei (figs. 35, 38)

Three peroneal muscles are differentiated, peronaeus longus, peronaeus brevis and peronaeus accessorius. Strong intermuscular septa intervene between the peronaei and neighboring groups.

Peronaeus longus is a well-marked fleshy belly arising from the lateral surface of the fibula from just below the head as far as the junction of upper and middle thirds of the shaft. It gradually narrows and gives place to a stout tendon which courses down the lateral surface of the leg, then inclines backwards to pass behind and below the lateral malleolus, being there bound down by a well-marked retinaculum. It proceeds round the tubercle on the lateral surface of the calcaneus and then enters the sole. In the sole it runs obliquely, enclosed in a strong fibrous sheath, to insert on the medial side of the base of the first metatarsal. No sesamoid was found in this tendon.

Peronaeus brevis lies beneath the tendon of the preceding as it occupies the distal half of the fibular shaft. Its tendon is somewhat slenderer, but rounded. It enters the foot lying posterior to the peronaeus longus tendon as it lies beneath the retinaculum. It terminates on the styloid process of the fifth metatarsal.

Peronaeus accessorius (= *peronaeus quinti digiti*) has an intermediate position at its origin. It springs from the middle two-fourths of the fibular shaft on a more posterior plane than the peronaeus brevis. Its tendon runs parallel to that of the previous muscle, but more posteriorly. It is somewhat more slender, but rounded, and inserts also in the fifth metatarsal.

h. Extensor Group of the Leg (fig. 38)

This group includes a large tibialis anterior, two long hallucial extensors and an extensor digitorum longus.

Tibialis anterior arises from the lateral surface of the tibial shaft from just below the head to about the mid-point of the shaft; and also from the interosseous membrane. It remains fleshy to about the junction of middle and lower one-third of the leg, then is replaced by a stout tendon which passes beneath the extensor retinaculum to the dorsum of the foot. Here it deviates towards the hallux and inserts on the base of its metatarsal.

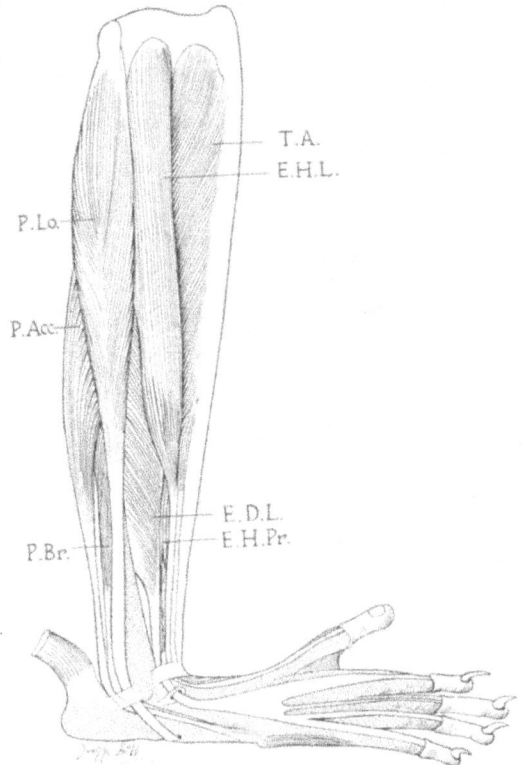

T.A.
E.H.L.
P.Lo
P.Acc
P.Br.
E.D.L.
E.H.Pr.

FIG. 38.   Dissection of the right crural region and dorsum of the foot.

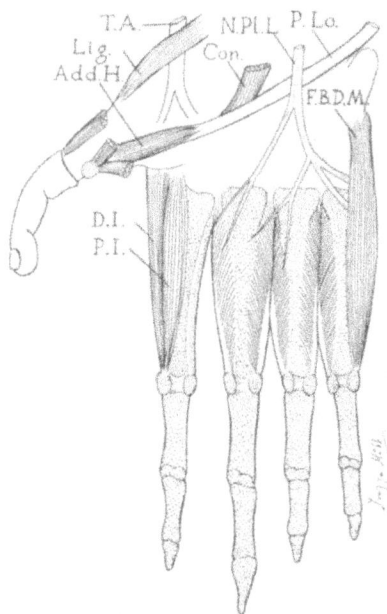

The main *extensor hallucis longus* is a long straplike fleshy band which springs mainly from aponeurosis over the uppermost part of the lateral half of the tibialis anterior, with some deep connections to the intermuscular septum between tibialis and extensor digitorum longus. It remains fleshy to the same distance as tibialis anterior, then gives rise to a stout rounded tendon which, after passing beneath the retinaculum lateral to the tibialis, proceeds parallel therewith to insert on the dorsal tendon expansion of the hallux, connecting thence with both phalanges. This muscle appears to correspond with the lateral portion of the tibialis anterior of Beattie's account of *Hapale*.

The long hallucial extensor is not the extensor tarsi described by Förster (1916) in *Hapale* and tentatively put forward as a homologue of the human peronaeus tertius. The latter is located on the fibular side of the limb and appears beneath the extensor retinaculum between the extensor hallucis longus and the extensor digitorum longus. The extensor hallucis of *Callimico* here under consideration corresponds rather with the extensor hallucis proprius of *Hapale* in Förster's account and also in that of Beattie. It is, as in *Hapale*, a thin slip arising deeply from the interosseous membrane and neighboring intermuscular septa, hidden for the most part by tibialis anterior. The tendon emerges just above the extensor retinaculum lateral to that of

the preceding muscle. It courses parallel with it and inserts on the base of the ungual phalanx of the hallux.

Extensor digitorum longus is a unipenniform muscle springing by fleshy fibres from the lower margin of the lateral tibial condyle and the whole length of the interosseous border of the fibula, as well as from the intermuscular septum between itself and the peronaei. A single stout tendon emerges just above the retinaculum and this remains single in the dorsum of the foot. Soon after emerging from the retinaculum it passes through a pulley-like fibrous structure attached to the lateral surface of the calcaneus just below the lateral malleolus. After this the tendon spreads fanwise to form a membranous sheet within which fibres concentrated in certain positions corresponding to the digital contributions but not fully differentiated from their matrix. Contributions are given to tendinous dorsal expansions of digits II, III, IV, and V.

Extensor digitorum brevis: This is thin, but well differentiated, supplying muscular bellies to digits II, III, and IV, but a tendinous slip only to V. There is no contribution to the hallux. The slip to the second toe is very obliquely disposed, the others progressively less so towards the fibular side.

SPLANCHNOLOGY

I. DIGESTIVE SYSTEM

1. BUCCAL CAVITY (fig. 40)

The thin lips are deeply pigmented externally and on their free borders, less so on the inner surface. Pigment extends on to the mucous membrane of the cheeks and gums as well as the palate, where it is more concentrated on the rugae than in the interspaces between them. The pigment decreases on the soft palate, but reappears slightly on its posterior edge (except in the central tubercle). The tongue is not pigmented, but the edge of the frenal lamella and its bifid apex are deeply pigmented.

a. Hard Palate

The hard palate is of oblong outline bordered in front by the incisors and laterally is parallel-sided by virtue of the arrangement of the canines and cheek-teeth. Between the canines and incisors each side is a broad diastema through which the palatal mucosa becomes continuous with that of the gums. The palatal mucosa is raised into six transverse rugae with the vestige of a seventh posteriorly between the two wisdom teeth. Except for the first, all these ridges are paired, being interrupted in the median line, the space between their medial ends being, in the case of the three middle pairs, partly filled by detached islands of raised mucous membrane. The foremost ridge is continuous across the median plane and is less distinctly raised than the others, being indicated rather by the pigmentary concentration than by its height. The second is composed

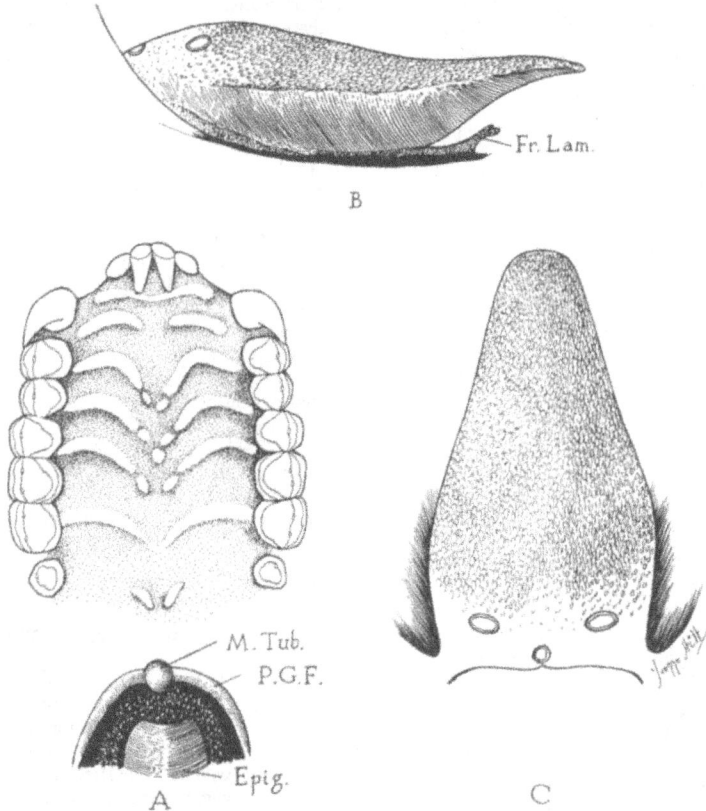

Fᵢɢ. 40.   A. Palate and upper dentition.   B. Tongue from the right side.
C. Tongue from above.

of two simple arches, convex rostrally, separated by a wide interval. The next three pairs are more angular, with the medial ends turned more acutely caudad. The sixth pair revert to the simple arch shape and have a lesser interval between their medial ends. These lie opposite M.^{2} Details are shown in figure 40.

The general picture thus conforms to the pattern demonstrated by Schultz (1949) in other hapalids, but with the additional extension of the rugae to the region between the extra molars. Phylogenetic reduction appears to have occurred in *Hapale* and to a further degree in *Leontocebus* and *Oedipomidas.*

b. Tongue (fig. 40, B, C)

In general outline the tongue conforms to the shape of the palate. Its narrow anterior free one-third is dorso-ventrally much compressed, slightly tapering forwards, and with a spatulate apex finely notched in the median line. The posterior half broadens rather ab-

ruptly, the maximum width, opposite the vallate papillae, reaching 11.5 mm. The dorso-ventral thickness of the organ likewise increases abruptly at the junction of the free anterior one-third and the fixed posterior two-thirds, the maximum for the former being 3 mm. while for the latter (from dorsum to level of buccal floor) it reaches 5 mm. The contour of the dorsum changes abruptly at the junction of the anterior (buccal) two-thirds and posterior (pharyngeal) one-third from horizontal to vertical by a rounded convexity. The length of the buccal portion from apex to posterior vallate papilla is 20 mm. From the center of this papilla to the base of the epiglottis the distance is only 2 mm., the median papilla being located on the pharyngeal aspect of the organ. The palato-glossal fold (anterior faucial pillar) gains its attachment to the tongue opposite the anterior (paired) vallate papillae.

As adumbrated above, there are three circumvallate papillae, arranged in a triangle, with the median member posteriorly. The median papilla is slightly smaller than the paired anterior ones. The top of each is flush with the general surface of the tongue, but the median one is slightly more prominent than the others. Each is surrounded by a deep, but narrow vallum. The anterior pair are located 6.0 mm. apart (measured between their centers), and each is 4.5 mm. from the median member.

The remainder of the dorsum linguae is uniformly clothed with a single type of papilla. This approaches the fungiform variety, but is not so large or prominent. There are no conical or filiform papillae. The papillae are broad topped, almost flattened like minute vallate papillae, and planted so as to render their free ends contiguous, giving a general roughened effect to the surface. Here and there are some slightly larger and more prominent individual papillae, but these have no special distribution or arrangement. Specialized comb-like apical papillae such as occur in *Hapale* are quite lacking.

Papillae cease abruptly at the edges of the dorsum both on the apical free part and on the body of the organ. The under surface of the free portion and the sides of the fixed part are covered with thin, smooth, tightly adherent mucous membrane. That on the sides exhibits a few feeble oblique sulci, but no specialized lateral organs.

The pharyngeal surface of the tongue faces directly backwards and is covered with smooth mucous membrane marked merely by slightly raised smooth, rounded discrete papillae uniformly scattered over its surface and well separated from each other.

Beneath the free anterior part of the tongue is a very distinct frenal lamella, produced forwards in an irregularly lobulated, deeply pigmented structure ending in a bifid apex. Laterally the lamella is continued along the line of attachment of the tongue to the floor of the mouth as a crenulated, pigmented fold as far back as the level of the anterior vallate papilla.

TABLE 5

DENTAL DIMENSIONS (IN MILLIMETERS)

Upper		Lower	
$I.^2$–$I.^2$	5 mm.	$I._2$–$I._2$	6.5 mm.
$I.^1$ crown length	2.2	$I._1$ crown length	3.0
crown breadth	2.0	crown breadth	1.0
$I.^2$ crown length	1.8	$I._2$ crown length	3.0
crown breadth	1.4	crown breadth	0.8
C. crown length	5.8	C. crown length	5.5
C. diameter at base	2.5		
Cheek-teeth, length	12.0	Cheek-teeth, length	14.0
Premolar length	5.8	Premolar length	6.7
Molar length	6.2	Molar length	7.3
$P.^1$ medio-lateral	2.7	$P._1$ medio-lateral	1.8
$P.^2$ medio-lateral	2.9	$P._2$ medio-lateral	1.9
$P.^3$ medio-lateral	3.0	$P._3$ medio-lateral	2.3
$M.^1$ medio-lateral	3.6	$M._1$ medio-lateral	2.4
$M.^2$ medio-lateral	3.3	$M._2$ medio-lateral	2.3
$M.^3$ medio-lateral	1.7	$M._3$ medio-lateral	1.75
Highest crown ($P.^3$)	1.8	Highest crown ($P._1$)	2.6

Between the frenal lamella and the under surface of the tongue in the median line is a distinct mucous fold or frenum. On the tongue this is continued as a faint ridge almost to the apical notch.

c. Dentition (fig. 40, A)

Dental formula: $I.\frac{2}{2}$, $C.\frac{1}{1}$, $P.\frac{3}{3}$, $M.\frac{3}{3} = 36$, i.e. as in Cebidae, the third molar, though small, being retained in both jaws, thereby differing from the Hapalidae. Some dimensions of the teeth are given in table 5.

Morphology of Individual Teeth

UPPER DENTITION: $I.^1$ crown with quadrate outline, the anterior aspect broad and spadelike, with horizontal free edge; palatal aspect oblique, the antero-posterior dimension thickening considerably from occlusal edge towards neck of tooth and marked by a triangular concave area, the long axis of which is directed downwards and medially, so that contact is made with corresponding point of fellow only at the lower medial angle of crown, a triangular interval separating the two teeth between this point and the alveolar border.

$I.^2$ much smaller, shorter and blunter than $I.^1$ and placed on a plane somewhat posterior thereto. When worn the crown is short and peglike, but unworn tooth with squat triangular crown, bulging considerably on both medial and distal sides immediately beyond neck, distal bulge separated from bluntly rounded apex by a notch, much as in *Leontocebus*. Relatively smaller than $I.^2$ of *Cebuella* and less conical than that of *Hapale*. In the two latter genera the medial and distal bulges are not seen. In *Cebuella* the occlusal edge is aligned with that of $I.^1$, but on a more posterior plane. *Leontocebus* is intermediate in this respect.

Upper canine: Crown almost thrice the length of that of incisors, from which it is separated by a narrow diastema. Conical with moderately sharp tip when unworn, roughly circular in cross-section at base, convex anteriorly and slightly concave behind, but hinder border sharp. Anterior surface smooth and convex, with a vertical groove along its medial part, extending from neck almost to apex, more marked than in *Hapale*, resembling rather *Cebuella*. A slight heel at base on postero-buccal side, and a slight cingulum on lingual side.

Upper premolars: Increasing in size from foremost to hindmost, two-rooted. All with bicuspid crowns, the buccal cusps much higher than the lingual and separated from them by deep medio-distal groove. Cingulum strongly developed on buccal side, especially in front, where an elevation almost amounting to a small cuspule is found. Buccal cingulum connected by sharper ridges to anterior and posterior edges of base of lingual cusp. These ridges form the boundaries of the medio-distal groove. Lingual surface of crown bulbous without cingulum.

Upper molars: Diminishing in size from M.^{1}–M.^{3}, three-rooted. Occlusal surface of crowns subtriangular as in Hapalidae, not tending to quadrangular form (owing to presence of large hypocones) met with in Cebidae (Dollman, 1937). In *Callimico,* however, hypocones are present though small, not affecting the general contour of the crown (*cf.,* however, Elliot, **3**: 261).

Buccal cusps higher than protocones, but paracones with broader bases than metacones. Protocones separated from buccal cusps by deep, broad medio-distal groove forming a fossa which receives, in occlusion, the postero-buccal cusp of the corresponding lower molar. A broad lingual cingulum on M.^{1} with its edge raised posteriorly to form small hypocone. Less developed on M.^{2} and lacking on M.^{3} Buccal cingulum narrow, but well defined on M.^{1} and with festooned free border suggestive of style-formation; at least a diminutive mesostyle present. These features are lacking on M.^{2} and M.^{3} M.^{3} is bicuspid, resembling a small premolar, with vestigial lingual cingulum and buccal cingulum lacking, the whole tooth being more reduced than M.^{2} of *Leontocebus.*

LOWER DENTITION: *Lower incisors* more vertically planted than in Hapalidae (Dollman) and not elongated to level of canines, thus resembling the tamarins (*Tamarin, Tamarinus, Leontocebus,* and *Oedipomidas*) and differing from *Hapale* (including *Mico* and *Cebuella*).

I.$_1$ with triangular spadelike crown, and with shovel-shaped posterior surface.

I.$_2$ similar but narrower, with lateral edge sloping upwards and medially from neck with a slight notch giving a rather falcate form (Ribeiro, 1940) near occlusal edge. Long axis of crown bent at an angle with that of root both medio-distally and antero-posteriorly.

Lower canine vertically planted and not separated from incisors or premolars by diastemata. Axis of crown bowed slightly forwards. A slight talon postero-basally formed by cingulum which passes thence forwards on lingual side of base, but not on to anterior or buccal surface. Three ridges descend from apex to neck, anterior, lingual, and intermediate, of which the last is most acute. The lingual (or posterior) ridge shuts off the talon from the rest of the tooth.

Lower premolars diminishing in height from foremost to hindmost, but crowns subequal in basal plan.

Foremost rather caniniform, its height but little less than canine, but talonid larger and prolonged on buccal side as sharp ridge along posterior edge of main cusp. The three other ridges present on canine are here present and even better marked, the anterior one being continuous with the anterior end of the lingual cingulum.

Middle premolar bicuspid with buccal and lingual cusps of equal height and joined by a ridge at their posterior ends, the ridge separating anterior and posterior foveae.

Hindmost premolar similar, but cusps lower and posterior fovea more expanded to form a rudimentary talonid basin.

Lower molars: Occlusal surface of crowns approximately quadrate with rounded angles, and the crown pattern much resembling that seen in *Cebus.*

Protoconids and metaconids opposite each other and subequal in height, connected by a transverse ridge separating an anterior fovea of diamond-shape from talonid basin. Fovea bounded anteriorly by thin enamel ridge. Talonid basin deeper than anterior fovea, bounded behind by thin enamel ridge which rises to high hypoconid on buccal side (separated by distinct notch continued as groove on buccal surface, from protoconid). Entoconid not present.

M.$_2$ slightly smaller than M.$_1$; M.$_3$ reduced still further and quadrate outline lost, the crown being rounded in basal plan, its cusps feebly marked and fossae reduced in area and depth.

d. Salivary Glands

Parotid and submandibular glands are about equal in size (*cf. Hapale,* Hill, 1957). The former is broad and short, embracing the external auditory meatus above, overlapping the masseter slightly in front, but failing below to make contact with the submandibular. Deeply it extends to the pharyngeal wall, enclosing the terminal part of the external carotid artery. Its duct, after the usual course across the masseter opens into the buccal vestibule opposite the hindmost upper premolar.

The submandibular is a compact ovoid lobulated mass partly hidden by the angle of the mandible. Its duct takes the usual course. The sublingual is a fusiform, finely-lobulated body, 9.3 mm. long, broader anteriorly than behind, beneath the mucosa of the buccal floor, lying parallel with the frenal lamella.

2. PHARYNX (figs. 20, 47B)

The nasopharynx comprises the elongated choanal passage (*vide* p. 107) which morphologically in Wood Jones' view (1940) pertains to the nasal fossa and not to the alimentary tract; it is therefore treated elsewhere. Its dorsal wall in the junctional region (nasopharyngeal isthmus) is lightly pigmented and highly vascular. Caudad of the level of the palatal velum pigmentation increases, affecting the whole dorsal wall of the oropharynx. In the laryngo-pharynx the pigment again diminishes, and only in this section of the tube is there any modification of the mucosa. Here a small, slightly elevated area of oval outline occupies the median plane opposite the aditus laryngis. It is delimited by a shallow sulcus and from the two lateral limbs of this sulcus shallower oblique sulci proceed laterally and caudally, gradually fading. Otherwise the mucosa of the dorsal wall of the pharynx is everywhere smooth.

The lateral wall of the oro-pharynx is occupied by the large tonsillar fossa, a depression of triangular outline with the apex at the edge of the soft palate. From the soft palate two strongly marked folds (faucial pillars) diverge as they descend. The anterior, 5.5 mm. long, proceeds downwards and forwards towards the junction of the buccal and pharyngeal parts of the tongue. The posterior proceeds more directly along the pharyngeal wall, fading gradually in the neighborhood of the laryngeal aditus. The former carries the palato-glossus and the latter the palato-pharyngeus muscle. The latter does not form a complete transverse bar. Near the apex of the triangular depression lies the pocket-like tonsil. This consists of a thickened J-shaped rim. The vertical limb of the J commences parallel with the anterior faucial pillar, but diverges from it slightly below. The upturned limb lies below and behind. A recess is formed which is deepest at the junction of the two limbs. Tonsillar follicles line the recess, which opens in an upward and backward direction. The recess measures 3 mm. high and 1.5 mm. broad.

Pharyngeal muscles are dealt with on p. 44.

3. OESOPHAGUS (fig. 49)

This tube commences at the lower border of the cricoid cartilage and terminates at the cardiac opening of the stomach, and is therefore divisible into cervical, thoracic, and abdominal parts.

Its total length is 75.5 mm., made up of cervical 33.5 mm., thoracic 36.5 mm., and abdominal 5.5 mm. From the dorsum linguae to the lower border of the cricoid the distance is 17.7 mm., which thus represents the length of the pharynx, making the total from the level of the floor of the buccal cavity to the stomach 93.2 mm. The transverse breadth of the upper oesophagus is 4.5 mm. and the diameter at the lower end 3.5 mm.

The tube takes an approximately median line course, but there is a slight diversion to the left in the lower neck followed by a dextral trend in the anterior thorax. At the level of the tracheal bifurcation the tube returns to the middle line, making a stronger leftward curve in the hinder thorax, but this is followed, before its passage through the diaphragm by a terminal dextral trend which is continued in the abdominal portion. The cervical part of the gullet is relatively thin-walled, pale in color and flattened dorso-ventrally. Commencing as a wide tract, it narrows gradually to a minimum at the thoracic inlet. There is no marked constriction at the level of the tracheal bifurcation, or at the oesophageal opening of the diaphragm. The thoracic portion is much thicker-walled, darker in color (probably from the greater quantity of venous blood), and circular in section. The transition is quite abrupt. Equally striking are the internal differences. The mucosa of the cervical segment is pale and provided with half a dozen delicate longitudinal rugae on both dorsal and ventral walls. These fade out before reaching the level of the thoracic inlet. The thoracic portion is provided with about 8–10 strongly marked broad-topped longitudinal rugae

which stand out very prominently from the walls, giving the lumen a stellate outline in transverse section. This appearance is continued in the abdominal part. The difference in thickness of wall seems to be due to the amount of tissue in the muscular coats of the tube.

In its last few mm. of the thoracic oesophagus possesses a triangular "mesentery" where the pleurae of the two sides are in contact between the dorsal wall of the tube and the thoracic aorta, the base of the triangle, of course, being formed by the diaphragmatic attachment.

The abdominal oesophagus widens somewhat in funnel-like fashion before opening into the stomach, but is not so large proportionately as shown in Wood Jones' figure of that of *Hapale penicillata* (Wood Jones, 1929). The tube is lodged in a deep notch on the dorsal margin of the liver, but is not so deeply embedded as in the lemurs and lorises.

4. STOMACH (figs. 41, 56)

In general shape this conforms fairly closely with that in marmosets, e.g. that of *Hapale penicillata* figured by Wood Jones (1929). In this type of viscus the cardiac opening is located approximately at the mid-point of the transverse dimension of the organ, this dimension lying almost at right angles to the longitudinal axis of the abdominal cavity. The lesser curvature is therefore short and equalled by the leftward projection of the fundus. The organ is thus a fairly regular pyriform shape, with the broad end to the left, forming the fundus, and narrowing quite regularly to the right towards the pylorus. The axis of the fundus is in line with that of the corpus ventriculi and not turned forwards as in *Hapale;* hence there is no

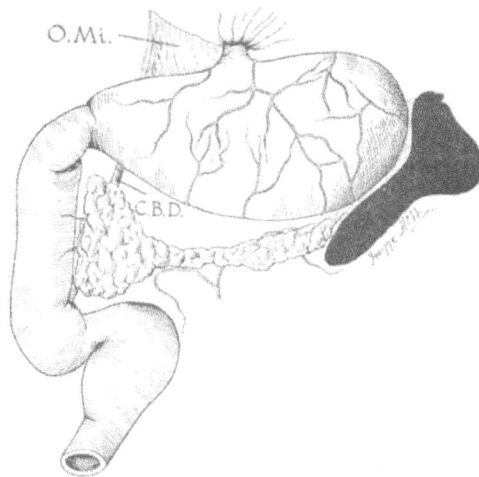

FIG. 41. Stomach, duodenum, pancreas, and spleen with adjacent vessels viewed from the ventral aspect.

cardio-fundic notch. The two main surfaces are not, however, located dorsally and ventrally, for the organ is rolled on its own axis in such fashion that the ventral wall looks more forwards than ventrally, the greater curvature therefore taking on a ventral position immediately caudad of the margin of the liver. The line of attachment of the great omentum lies still further caudad (fig. 41). Moreover, prior to disturbance the long axis is somewhat bowed towards the pyloric end, the pylorus being somewhat dorsal in position, so as to be hidden ventrally by the pyloric antrum.

Only vaguely can the organ be divided into fundus, corpus, and pyloric segment, for there is no distinct incisura angularis or sulcus intermedius; but the fundus in any case accounts for the largest part of the viscus.

Internally the fundus region is smooth. From the cardiac orifice two raised folds course along the lesser curvature, enclosing a groove between them. The dorsal fold is longer than the ventral, proceeding as far as the pylorus. A third minor fold lies between the two main ones.

The transverse length of the stomach, moderately filled, is 46 mm., of which the fundus accounts for 24 mm. The diameter of the gastric tube reaches 24 mm. At the pyloric constriction the diameter is 14.5 mm.

5. SMALL INTESTINE

This is a fairly wide tube, but narrows slightly in certain parts. It is divisible into duodenum and jejuno-ileum.

The duodenum is relatively short but wide and capacious. There is no duodenal "cap" immediately following the pyloric constriction, the gut coursing caudad almost at once. Hence there is strictly no "first" part like that in Man, but a "second" part some 33 mm. long and 17.75 mm. in diameter. This lies parallel to but to the right of the vertebral column. It terminates posteriorly opposite the body of the third lumbar vertebra by turning abruptly craniad and then almost immediately caudad again, forming thus an S-shaped bend, of which the second component is the duodeno-jejunal flexure. This flexure is very bulbous, narrowing rapidly thereafter in the anterior part of the jejuno-ileum. The duodenum is covered by peritoneum on both surfaces, but its mesentery has been lost by adhesion to the dorsal parietes. The tube is further anchored by raised peritoneal bands, one connecting the convexity of the first bend to the mesentery of the descending colon (recto-duodenal fold of Klaatsch) and the other from the concavity of the same flexure on to the ventral surface of the pancreas.

Internally the duodenum presents a fairly smooth appearance. The mucosa is thin but very vascular and pancreatic lobules may be outlined through the sinistral wall of the gut, each surrounded by injected arteries. The opening of the conjoined ducts from the

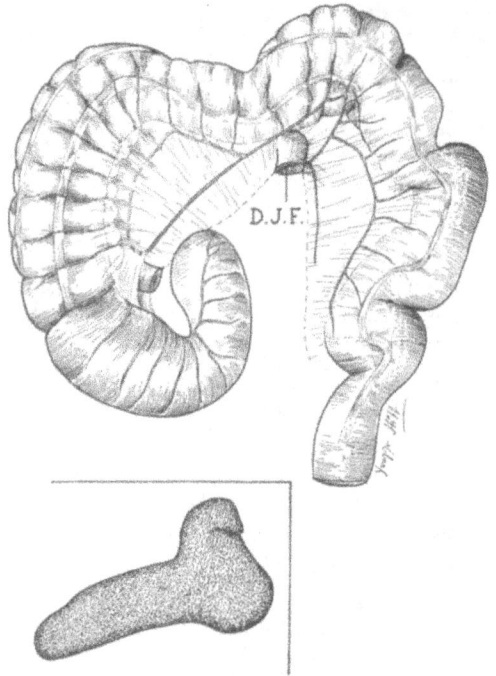

FIG. 42. Disposition of the large intestine and its mesenteries and vessels. Inset shows the spleen from its parietal aspect.

liver and pancreas is situated 10 mm. beyond the pylorus on the dorso-medial wall of the duodenum. It is very inconspicuous, having no papilla or other associated modification of the mucous lining. It is surrounded by pancreatic lobules just prior to its entry through the gut wall.

The jejuno-ileum presents little for remark. It is relatively short, though longer proportionally than in *Tarsius* (Hill, 1955), being thrown into six major loops by virtue of the laxity of the mesentery. From the bulbous duodeno-jejunal flexure the tube narrows rather abruptly to a diameter of 11.5 mm., increasing later to 14.2 mm. at about the middle of its course and still further before ending at the ileo-colic junction, where its diameter is 17.0 mm., or almost as great as the duodenum. Internally mucous folds are absent, but the usual villous appearance is seen. No macroscopic lymphoid patches were noted.

6. LARGE INTESTINE (fig. 42)

The large bowel takes the usual horse-shoe-shaped course around the periphery of the coils of the jejuno-ileum, with the caecum forming a bent sac to the pelvic side of the ilio-colic junction—i.e. as in the Hapalidae, *Aotes, Callicebus,* and some other genera of Platyrrhini.

The gut is provided with a mesocolon throughout its course, but this is extremely short at the point where the transverse colon crosses the vertebral column—a primitive mammalian feature. On either side of this point, however, the mesocolon is very lax, and particularly on the left, where the descending colon is thrown into a number of secondary loops.

The total length of the large intestine from the apex of the caecum to the anus, measured along the greater curvature, is approximately 413 mm.

The diameter varies considerably, but is everywhere greater than the average for the small intestine. Table 6 gives the dimensions at different levels.

TABLE 6

	Diameter in millimeters
Basal caecal sac	23.2
Apical region of caecum	16.8
Proximal colon	32
Transverse colon	21
Distal colon	17
Anterior part of rectum	12.5
Terminal part of rectum	9–10

The ileum enters the colon almost at right angles, but with a slight craniad inclination. Opposite the junction the gut shows a well-marked annular sulcus, sharply demarcating the basal sac of the caecum from the ascending colon, indicating the site of a caeco-colic sphincter such as Johnston (1920) described for *Callicebus*. Caudal to this the caecum turns slightly towards the median plane for a little over half its length, then abruptly craniad dorsal to the terminal ileum, to end in a blunt, broadly rounded extremity. The caecum is not tethered to the ileum by a median bloodless mesotyphlon, but is provided with dorsal and ventral symmetrical vascular folds, which contain fat and convey vessels to the caecal wall. *Hapale, Tamarin, Leontocebus,* and *Oedipomidas* possess the anangious intermediate fold (Hill and Rewell, 1948). The caecum differs from the colon in the smoothness of its wall.

Internally the caecum presents a velvety appearance, but shows no obvious lymphoid patches. The ileocolic opening is very large, 3.8 mm. across, subcircular in outline and guarded by a raised thickening of the mucosa which increases the diameter to 6 mm. There is no clear telescoping into the caecal lumen or any specialized folding of the neighboring mucosa. The colon is distinctly sacculated, but the haustra so formed are not as sharply defined as in catarrhine monkeys. They are of the same order as is commonly, but not invariably found in marmosets and tamarins, but they are not effaced by distension of the gut. This is because the whole length of the colon is provided with three well-defined longitudinal muscular bands (*taeniae coli*). These taeniae are equidistant from each other, but are located differently from their counterparts in man and the higher Primates, being therefore analogous

but not homologous therewith. They do, however, correspond with those observed in, e.g., *Hapale jacchus* and *Cebuella pygmaea*. On the transverse colon the taeniae are arranged one ventrally, one along the cranial wall, and the third along the caudal wall of the tube. None of them is mesocolic, the mesocolon being attached dorsally between the cranial and caudal taeniae. Traced on to the ascending colon the ventral taenia becomes lateral, while the cranial becomes dorso-medial and the caudal ventro-medial in position. Continued on to the descending colon the ventral taenia again becomes laterally placed, the cranial dorso-medial and the caudal ventro-medial. In contrast to *Hapale* (Hill and Rewell, 1948) the taeniae are not continued on to the basal caecal sac.

Whereas in *Hapale* the great omentum gains no attachment to the colon, but is adherent to the right half only of the transverse mesocolon, *Callimico* is similar, but a few raised folds from the omento-mesocolic fusion are carried on to the colic wall. These peritoneal foldings recall those connecting the limbs of the colic ansa of *Hapalemur* (Davies and Hill, 1954) but are much more delicate and, of course, less numerous. The right mesocolon is fat laden, but the fat does not form a continuous stratum; it is located near the root of the membrane, with prolongations along the branches of the colic arteries as far as the margin of the gut. Between the terminal vessels are fat-free fenestrae (*cf. Tamarin midas* as described by Hill and Rewell, 1948). The base of the right leaflet of the left mesocolon anteriorly is connected with the duodeno-jejunal flexure by a raised peritoneal fold. This is additional to the recto-duodenal fold mentioned above.

The rectum (fig. 51) shows no sharp division from the colon, and is of similar caliber. Rather thin-walled, it continues straight back along the dorsal wall of the pelvis without sagittal or lateral flexures. The mucosa is quite smooth. At its junction with the anal canal an ampulla is formed by virtue of an infolding of the wall affecting mucous, submucous, and muscular coats. The depression on the oral side of this infolding is particularly deep on the mid-dorsal line.

The anal canal accounts for the terminal 5 mm. of the alimentary tube. It is demarcated from the rectum by the afore-mentioned infolding of the wall. This projects some 5.5 mm. into the lumen, forming during the resting stage a valvelike occlusive device. It probably represents the thin shelflike structure described in some species of *Lemur* (Hill, 1958) but is here more robust without the membranous character.

7. LIVER (figs. 43, 44, 45)

The organ weighs 8.7 gm. and measures 41 mm. transversely, 32.5 mm. in the cranio-caudal dimension and 37 mm. dorso-ventrally.

In general shape it agrees with that of other generalized platyrrhines, presenting a deep notch dorsally be-

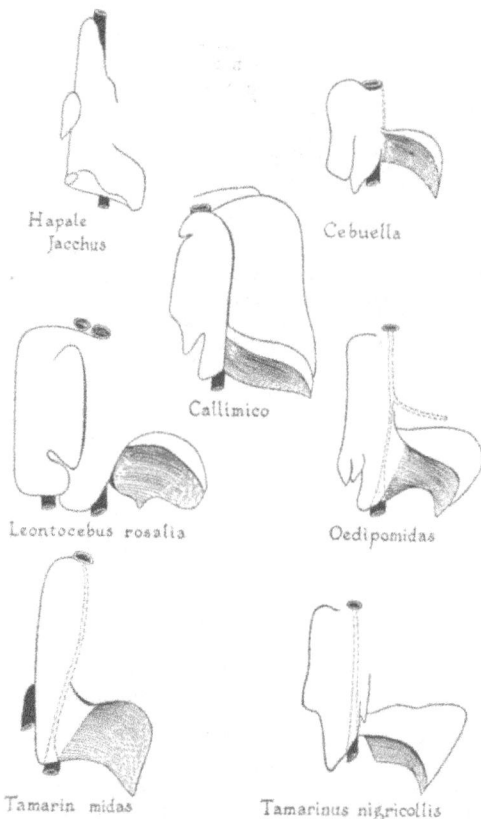

FIG. 43. Diagrams of the spigelio-caudate complex and related
structures in the livers of a series of Hapalidae.

gards the right and caudate lobes). The two contrasting forms both occur within the Hapalidae. *Hapale*
shows the elongated form, with maximum contrast between cranio-caudal length of right and left halves.
Oedipomidas spixi has the liver of the same general
form as in *Callimico; Tamarin midas* and *Tamarinus
nigricollis* are intermediate, the former inclining more
towards the hapaline type than the latter. *Leontocebus*
is also rather intermediate, but proportionate to body
size is nearer to *Hapale* than to *Tamarin*.

The organ is divided into three main lobes, left,
central, and right, by deep fissures virtually penetrating
the whole thickness of the liver substance. The central
lobe is further subdivided, but only topographically,
by the attachment of the falciform ligament; while
from its dorsal surface springs the large caudate lobe,
which proceeds across the visceral aspect of the right
lobe, where it contributes to the dorsal aspect of the
gland beyond the caudal margin of the right lobe.

The largest lobe is the central lobe but it exceeds the
left and right lobes only by a slight amount. If the
caudate lobe is included the right lobe exceeds the
others.

The central lobe overlaps the right and left lobes
ventrally, obscuring them from view in the undisturbed
state of the viscera, and it also contributes the largest
share to the diaphragmatic surface, but the discrepancy
here is very slight, for all three lobes contribute almost
equally to this surface. The liver of *Leontocebus* in
contrast has a small central lobe exceeded in size by
right and left lobes both of which enclose its caudo-
ventral margin by approaching together.

tween its right and left halves, but this is not so narrow
and acute as in the prosimian liver. Ventrally it is
highly convex in the transverse dimension, conforming
to the shape of the abdominal parietes. The extensive
cranial aspect is much less convex, but presents two
minor elevations separated by a median (cardiac)
depression, correlated with the cupolae of the diaphragm. The dorsal and caudal (visceral) aspects are
highly irregular from molding upon adjacent viscera.

As a whole the liver of *Callimico* resembles that of a
cebid monkey more especially *Aotes* or *Callicebus* than
that of *Hapale* as described and figured by Beattie.
There is less asymmetry between the amount of hepatic
tissue on either side the dorsal incisure, while the sub-
phrenic contour is less evenly domed, though this must
vary somewhat according to the state of contraction or
relaxation of the diaphragm.

The liver is of the broad, squat type rather than
the cranio-caudally elongated form (especially as re-

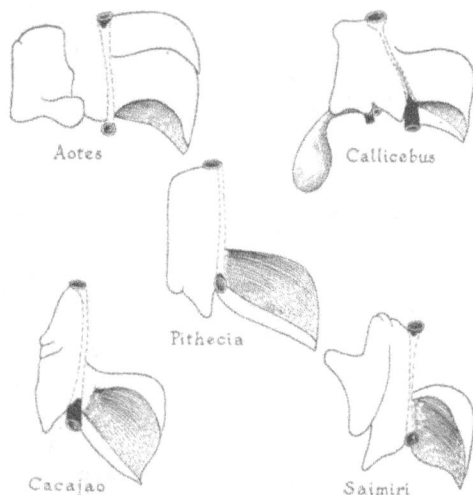

FIG. 44. Diagrams of the spigelio-caudate complex in a series
of cebid monkeys, for comparison with figure 43.

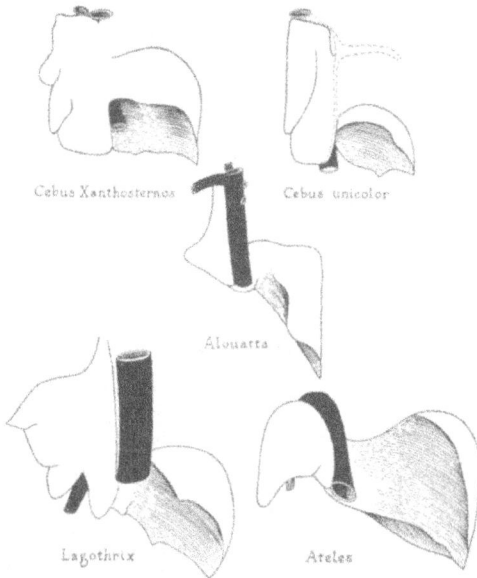

Fig. 45. Diagrams of the spigelio-caudate complex in a further series of cebid monkeys, for comparison with figure 43.

The umbilical notch, very shallow, divides the central lobe into a smaller left and larger right half, the division being continued on the diaphragmatic surface by the line of attachment of the falciform ligament. *Oedipomidas*, whose liver of all the hapalids most resembles that of *Callimico* in other respects, has a deep umbilical fissure cutting half-way through the thickness of the lobe both on parietal and visceral surfaces. In *Hapale* there is a broad, shallow notch confined to the caudo-ventral margin. *Tamarinus* resembles *Oedipomidas* in respect of this fissure, but it is not so deep, while *Leontocebus* agrees in having a definite fissure extending to the diaphragmatic surface. *Tamarin midas* is the most advanced, having a deep Y-shaped umbilical fissure, the left limb of the Y cutting almost horizontally into the left half of the central lobe.

On the visceral aspect of the central lobe flush with the edge of the lobe is located the gall bladder, an elongated pyriform sac, with its fundus narrowing to a point whence a membranous fold, derived from the serous investment of the sac, is continued peripherally halfway to the ventral border. The fold there takes on a right-angled bend, still attached to the liver tissue, and ends by joining the serous covering of the ligamentum teres. The fold is accompanied by a small artery, a continuation of the cystic artery. From the neck of the gall bladder the wide cystic duct proceeds dorsad to the porta hepatis, whence receiving the hepatic ducts, it continues as the common bile duct,

first within the lesser omentum, thereafter dorsal to the line of attachment of the head of the pancreas to duodenum, opening finally into the latter as already indicated (p. 68).

The Spigelian "lobe" is an elongated, oblong area of liver tissue on the dorsal aspect of the viscus adjacent to the right lobe, but facing somewhat medially, i.e. towards the great dorsal notch. Its cranial end has a rounded outline, but its two long sides are parallel, the right one being separated from the right lobe by a shallow groove only. This groove marks the site of the postcaval vein which courses through the liver substance and is exposed nowhere. Caudally this lobe ends in the two tongue-shaped processes, a shorter medioventral (or papillary) lobule and a longer dorso-lateral or caval lobule, separated by a V-shaped notch. From the right border of this caval lobule and from the hepatic tissue on the ventral side of the postcaval vein arises the caudate lobe. The peduncle of the caudate lobe is relatively restricted compared with the arrangements in some other platyrrhine monkeys. The lobe is dorsoventrally compressed, and excavated on its caudal aspect. It is almost entirely hidden dorsally and on the right by the overlapping margin of the right lobe.

The Spigelio-caudate complex is of some interest from the comparative aspect in so far as a wide range of variation occurs among the different genera of Platyrrhini. These morphological differences would appear to be of some significance from the genetic point of view in so far as they can hardly be influenced by environmental conditions. The condition in *Callimico* should therefore be of some value in assessing its systematic status relative to the Hapalidae and Cebidae. The problem is not easily solved for the variations are great, and so far we do not know sufficient about the range of individual variation. That this is not likely to be very great, however, is supported by the fact that in a series of five livers of *Lagothrix lagotricha* no substantial individual difference was encountered in respect of the features under consideration. The findings are listed below *seriatim*, and illustrated in figures 43–45.

I. Hapalidae

1. *Hapale jacchus.* Spigelian area narrow, elongated, narrowing anteriorly to rounded apex. Medial border rectilinear with a small isolated lobule projecting from it about its middle (not constant, ? papillary lobule). Lateral border ill defined, not demarcated by groove, except cranially, leaving broad union with right lobe over buried postcaval vein. Caudate lobe short, broad antero-posteriorly, with wide peduncle. No separate caval lobule which is represented merely by medial continuation of caudate tissue over dorsum of caval vein. Caudate almost entirely obscured dorsally and laterally by overlapping edge of right lobe (as in *Callimico*).

2. *Cebuella pygmaea.* Spigelian area oblong, broader and shorter than in *Hapale* and not narrowed anteriorly. Anterior border convex, ending laterally in broad area of contact with right lobe, without groove or fissure separat-

ing them. Medial border smooth, slightly concave. Posteriorly there are two distinct tongue-shaped lobules separated over wide extent by deep fissure opening behind into V-shaped notch. Medial (papillary) lobule squared, with rounded angles; caval lobule longer and more pointed. Caudate lobe with broad peduncle, obscured dorsally by right lobe, but projecting as a triangular surface beyond it laterally.

3. *Tamarinus nigricollis.* Spigelian area a narrow oblong produced posteriorly into a conical flattened lappet to the medial side of the dorsal wall of the postcaval vein; a minute secondary lappet arising from the base of the main lappet on its right side and separated by a short fissure. To the left of the base of the main lappet a small square shoulder of liver tissue demarcates the lappet from the medial margin of the Spigelian area. Right border of Spigelian unmarked, except in caudal one-third, where a distinct sulcus separates it from right lobe. Postcava entering near base of main lappet on its right side, whereas lesser lappet is to right of this. The identity of the main lappet is doubtful; it may represent a large caval lobule, in which event the secondary lappet is an additional process and the left shoulder represents the papillary lobule. Or does the main lappet constitute an enlarged papillary and the secondary one a vestigial caval lobule?

Judging by the position of the mesohepar (ligamentum hepato-cavo-phrenicum of Ruge, 1902), which proceeds in a slight groove along the dorso-lateral edge of the Spigelian mass thence to the right border of the principal lappet to end around the post-caval, the first view seems to be the correct interpretation.

Caudate lobe with narrow peduncle, but broadening rapidly to right. The broad dorsal aspect is marked by the right lobe, but a large area is exposed dextrally. The caudo-ventral border of this lobe presents a secondary apex near the peduncle.

4. *Tamarin midas.* Spigelian area a very narrow elongated oblong, with truncated anterior border, rectilinear left border and no demarcation whatever from right lobe dextrally. Posteriorly a deep fissure occurs between the right lobe and the root of the caudate. Medial to the caudate peduncle the Spigelian ends in a single convexity not prolonged as a lappet. There is no caval lobule or papillary process. The whole thickness of the posterior end of the Spigelian mass turns aside to form the caudate peduncle. Altogether the simplest arrangement met with.

5. *Oedipomidas spixi.* Spigelian area oblong, but less elongated than in *T. midas.* Cranial border ill-defined; medial border rectilinear. Lateral border demarcated only by line of mesohepar, which gives off dextrally a "coronary" ligament to dorsum of right lobe. Posteriorly a short bifid papillary process to left and an elongated lappet over postcava separated by deep fissure on right from peduncle of caudate. Caudate peduncle broad and thick. Caudate not entirely masked by right lobe either dorsally or on right.

6. *Leontocebus rosalia.* More complex than any other hapalid liver. Spigelian area oblong, with rounded corners, broader proportionally than in *Tamarin.* Anterior border slightly convex, medial border rectilinear. Lateral border demarcated from right lobe by distinct groove, except at anterior end where the groove arches over, sinistrally cutting into Spigelian tissue. Posteriorly are large papillary and caval lobules. The former bifid, but the left apex larger and curving round the smaller right, the two separated by an oblique incisura. Left papillary process with a right angled left border. Caval lobule tonguelike, but broad, slightly longer than combined papillaries.

Caudate peduncle broad and thick. Caudate relatively small, hidden dorsally by right lobe, which descends far beyond it. Caudate visible dextrally only as a small triangular mass in V-shaped notch between dual apices of right lobe. Base of caudate, however, having extensive sinistral surface passing beyond Spigelian complex both ventrad and caudad, facing directly medially. Ventral border with short, triangular process near peduncle, as in *Oedipomidas.*

II. Cebidae (figs. 44, 45)

1. *Callicebus cupreus.* Spigelian area short and broad, quadrate with rounded angles; medial border concave, cranial border slightly concave; no separation by groove or fissure from right lobe, but a deep incisure posteriorly separates it from neck of caudate lobe. Caudal border produced into three processes, two small papillary lobules on left separated by shallow concavity and a larger pointed caval lobule projecting over left wall of postcava. Rest of postcava buried in hepatic tissue. Caudate short and thick, broadly exposed on dorsal and lateral sides.

2. *Aotes trivirgatus griseimembra.* Spigelian area much as in *Callicebus*, as broad as long and facing almost directly mediad, but medial border almost rectilinear and cranial border somewhat convex. Caval lobule short, broad with rounded convex edge, to left of which the Spigelian terminates in a concave free border. A deep fissure cuts into the right (topographically dorsal) border of the Spigelian area cutting off a broad tongue of liver tissue which slightly resembles the sinistral papillary lobule of the liver of *Leontecebus.* Postcava completely buried in hepatic tissue; the attachment of the mesohepar connecting the cranial and caudal openings of the vein. Caudate as in *Callicebus* with even wider area of exposure laterally and dorsally.

3. *Pithecia monachus.* Spigelian area oblong, craniocaudally elongated, transversely narrowed. Cranial, medial, and lateral borders rectilinear. Caudal border with small rounded papillary lappet separated by V-shaped notch from longer, triangular pointed caval lobule. Postcava as in *Aotes.* Caudate large and exposed widely on right and dorsally, the dorsal exposure confined however to the large concavity related to the right kidney and adrenal.

4. *Cacajao rubicundus.* Spigelian area unique in its elongated, narrow almost fusiform outline, the cranial end narrowed, but rounded, the caudal end formed by a pointed caval lobule involving the whole width of the area. Medial border with several small notches leading to grooves. Presumably these represent displaced papillary processes. Postcava buried. Caudate as in *Cebus*, relatively small, not obscured dorsally by right lobe, but overshadowed elsewhere; peduncle thick.

5. *Saimiri sciureus.* Spigelian area broader than in *Cacajao*, but resembling it in medial position of papillary components. Two papillary lappets present of relatively large size, one projecting as a finger-like process from middle of medial border, the other more rounded and flaplike projecting caudo-mediad, and separated by deep V-shaped notch (continued on to liver substance as a fissure) from the sharply pointed caval lobule. Postcava entirely buried, but line of burial marked by a fissure, hidden beneath attachment of mesohepar. Caudate short and thick, its excavated dorsal surface freely exposed, and right aspect concave forming part of renal impression, which is completed by the small right lobe.

6. *Cebus apella xanthosternos.* General arrangement much as in *Saimiri*, but left border of Spigelian area with

two distinct papillary lappets one above the other, both directed caudo-medially. Remainder of posterior end of Spigelian forming a broad flat area not specially forming a caval lobule but with its right posterior angle projected over dorsal wall of postcava. Postcava, as found by Ruge in *C. capucinus*, buried in liver tissue and with same relations to ligamentum hepato-cavo-phrenicum. Caudate lobe enlarged in cranio-caudal dimension, but short transversely, with thick, wide peduncle, completely exposed dorsally having displaced right lobe almost entirely ventrad so that only the diaphragmatic portion of the latter is visible on dorsal aspect. Dorsal aspect of caudal lobe divided by ridge into convex area (related to parietes) and a concavity (related to kidney).

7. *Cebus unicolor.* Spigelian surface oblong and divided by a deep oblique fissure into antero-medial and posteromedial areas. The fissure cuts the medial margin at junction of middle and posterior thirds, the antero-medial area thus representing a single papillary area, the larger posterolateral district being the caval lobule which resembles that of *C. apella*, but is narrower transversely. Postcaval vein as in *C. apella*. Caudate relatively smaller and much less exposed, only a small triangular area just dextral to postcava exposed dorsally, the rest being overshadowed by right lobe.

8. *Alouatta belzebul.* Spigelian area reduced, of triangular form recalling that of *Hapale.* All three angles rounded, the anterior long and narrow (Spigelian proper). Left broad and short (papillary) and pointing directly mediad and the postero-dextral lappet-like related to postcaval wall. Postcava entirely exposed dorsally as in *Ateles* (Ruge) and *Lagothrix*, but ventrally, near its first contact with liver tissue, a folded stratum of hepatic tissue wraps over the wall to connect the caval lobule with the peduncle of the caudate lobe. Caudate pyramidal with apex at peduncle and base to right; greatly exposed dorsally as in *Cebus apella* by displacement of right lobe ventrad. A medial extension from caudate covering part of visceral aspect of right lobe.

9. *Lagothrix lagotricha.* Spigelian area extensive, though narrowed anteriorly to a sharp apex much as in *Alouatta.* Extension due to sinistral projections along medial margin —an anterior and a posterior papillary lappet separated by an oblique smooth edge. Posterior papillary lappet separated from triangular caval lobule by V-shaped notch. Postcaval vein exposed throughout dorsally as in *Alouatta* and *Ateles.* Caudate large, but not so extremely hypertrophied as in *Ateles;* its peduncle thick and obliquely attached to Spigelian base.

10. *Ateles paniscus.* Spigelian area with arc-like anterior edge, the right and left borders diverging behind; the former the longer owing to caudad projection of a single large papillary lappet. Caval lobule small and separated by wide interval from papillary process and also by a short fissure at left side of its root. Postcava entirely exposed dorsally. Caudate peduncle broad in sagittal diameter and the lobe enormous in size, exceeding the right lobe, which, however, it does not entirely shut off from the dorsal aspect. Appearances agree in all particulars with those recorded by Ruge for *A. ater.*

The significance of the above findings will be considered in the discussion.

8. PANCREAS (fig. 41)

This calls for little remark, resembling that in Hapalidae. It consists of an attenuated stratum of lobules each somewhat elongated in the sagittal diameter and narrow transversely, held together by loose areolar tissue. The borders of the gland are ill defined. It extends from the left wall of the duodenum to the hilum of the spleen and is wholly retroperitoneal. A single duct is present which joins the common bile duct immediately prior to the perforation by the latter of the duodenal wall. A large branched duct drains the "head" of the pancreas into the main duct just before the latter unites with the bile duct. Another large duct opens on the opposite (i.e. cranial) wall of the main duct. The union of these various ducts is surrounded by vascular channels which also form a collar round the final ampulla as it enters the duodenal wall.

II. RESPIRATORY TRACT

1. LARYNX (figs. 46, 47)

This is of the usual platyrrhine type (Lampert, 1926) as viewed from the pharynx. From the epiglottis to the lower border of the cricoid the distance is 12.2 mm (or 17.7 mm. from the dorsum linguae to the same level).

The cavity closely resembles that of *Hapale jacchus* as described by Beattie, being limited above by the epiglottis, aryteno-epiglottidean folds, and the mucous membrane covering the prominences formed by the tops of the arytenoids. Ventral to the last-mentioned are two further prominences enclosed within the arytenoepiglottidean folds and lateral margins of the base of the epiglottis. These are due to the large cartilages of Wrisberg (cuneiform cartilages). Between the two arytenoids the mucosa dips down in a deep notch, the bottom of which is 5 mm. caudal to the upper margin of the epiglottis.

From the aditus the cavity narrows rapidly to the level of the vestibular folds (false vocal cords) which are thick and prominent. The thinner true vocal cords are visible from above only on separating the upper folds. A distinct shallow ventricle intervenes between the false and true cords on each side. This proceeds laterally and dorsally beneath the corresponding vestibular fold as far as the free border of the thyroid ala.

a. Cartilages of the Larynx

The epiglottis is somewhat quadrangular in outline, 5 mm. across, but narrower caudally. Its free edge is thickened and fails to reach the soft palate above. It is feebly notched in the median line and its upper lateral angles are rounded. Neither median nor lateral glosso-epiglottic folds are present.

The thyroid cartilage is composed of two thin quadrangular plates somewhat thicker as they approach the median line ventrally where they unite at an angle of about 60°. The cranial border is connected to the hyoid by a short thyro-hyoid membrane, most of which is hidden by the overhang of the hyoid base. Dorsally,

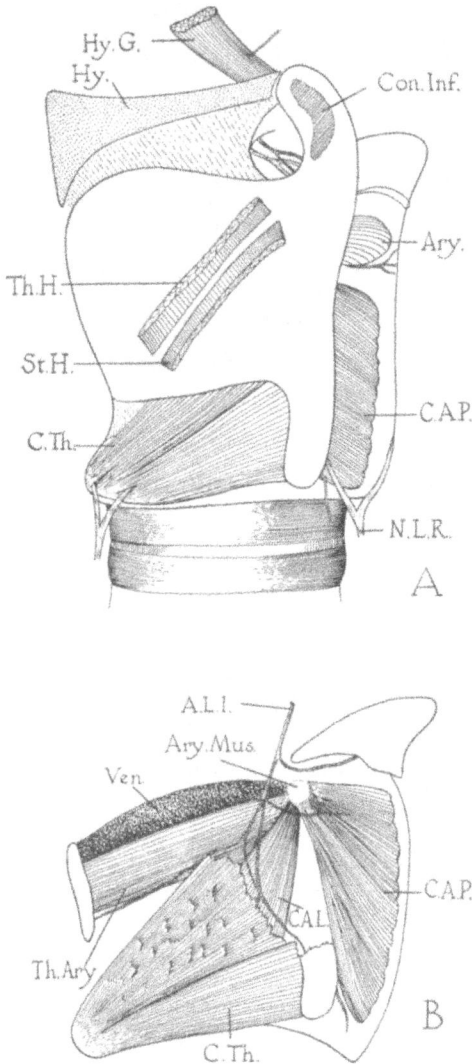

FIG. 46. Dissections of the larynx. A. Superficial dissection from the left side. B. Deep dissection from the right after removal of the ala of the thyroid cartilage.

however, the border is prolonged cranially as a long, blunt superior cornu, which curves ventrad as it proceeds and finally articulates with the great cornu of the hyoid by a mobile joint. The inferior pharyngeal constrictor muscle is attached to the joint and also to the hyoid and superior thyroid cornu in relation thereto. A similarly shaped cornu projects caudally from the postero-dorsal angle of the thyroid ala; it is straighter

than the superior cornu, but slightly curved. Its deep aspect forms a mobile joint with the edge of the cricoid. The dorsal border of the thyroid presents, as a result of the shape of its cornua, a sinuous outline. The oblique line is present on the external aspect of each ala.

The cricoid has the usual signet-ringed shape, with the cranial border sloping gradually from the expanded dorsal lamina on to the thin ventral part of the ring. The expanded part is marked by a strong longitudinal median ridge separating two lateral concavities occupied by the origins of the two dorsal crico-arytenoid muscles. Another ridge or thickening affects the cranial border of the dorsal expansion, and upon this are situated facets for the bases of the arytenoids.

The arytenoids are of the usual shape, but relatively large for the size of the larynx. Lampert (1926) states that the axis of the crico-arytenoid joint is horizontal in *Hapale,* but Beattie found it to be inclined slightly upwards and medially. In *Callimico* the extensive facet on the arytenoid is located on the caudal aspect of the base of the muscular process facing caudad and distinctly medially. The vocal process is small. Towards the pharynx the arytenoid is prolonged into a hook-shaped, dorsally directed, process covered ventrally, laterally and medially by mucous membrane, but occupied dorsally by a thick, muscular mass connecting the two arytenoids together.

b. Ligaments of the Larynx

The thyro-hyoid ligament is very restricted in view of the high position of the thyroid, whose cranial border is flush with the caudal margin of the hyoid—except dorsally, where the great cornu recedes and a triangular space is formed, guarded dorsally by the superior cornu of the thyroid. The membrane is here perforated by the anterior laryngeal vessels and nerve. The cranial margin of the thyro-hyoid ligament or membrane is attached to the dorsal surface of the hyoid, leaving a large bursa between itself and the caudal part of the bone.

The crico-thyroid ligament takes the usual form, with a strong median ventral portion connecting the two cartilages, and lateral portions, providing the basis for the conus elasticus (crico-vocal membrane). The latter extends forwards and medially from the edge of the cricoid and becomes separated from the thyroid ala by a stratum of muscle fibres. The cranial edge of this portion constitutes the true vocal ligament. Dorsally the crico-thyroid membrane contributes the capsular fibres to the crico-thyroid joint between the inferior cornu and the cricoid.

The crico-arytenoid joint is provided with a lax capsular ligament.

c. Muscles of the Larynx (figs. 46, A, B; 47, A)

The oblique line of the thyroid cartilage receives the attachments of the sterno-thyroid and thyro-hyoid

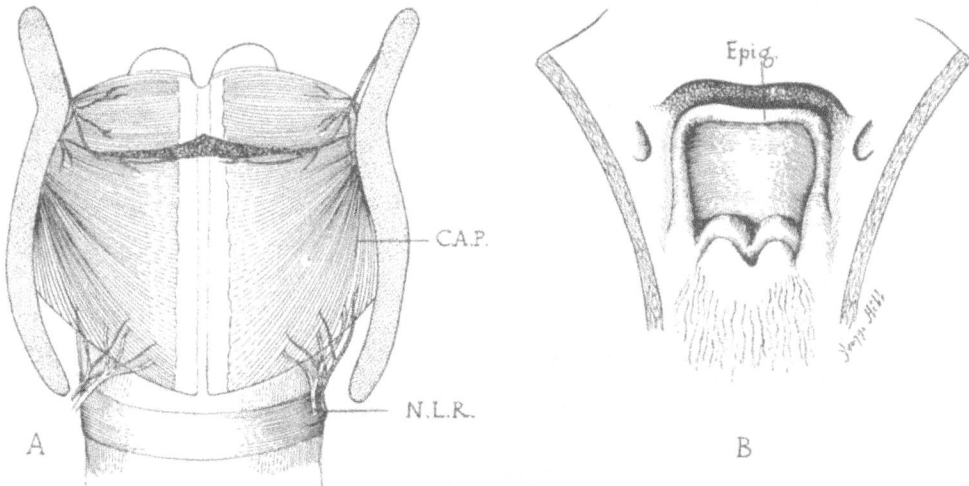

FIG. 47.　Further dissections of the larynx.　A. Dissection from the dorsal aspect.
B. View of the aditus laryngis from the laryngopharynx.

muscles.　The inferior pharyngeal constrictor has an origin from the superior cornu of the thyroid.

The *crico-thyroideus* consists of two parts fairly completely differentiated from each other, the two comprising a fan-shaped mass arising from the median ventral part of the cricoid, the muscles of the two sides diverging, but leaving little or no exposed crico-thyroid membrane between them.　The longest fibres insert on the ventral edge of the inferior thyroid cornu and adjacent part of the posterior border of the ala.　The remaining fibres insert on the posterior border of the ala more ventrally, many of them passing craniad deep to the ala to insert on its inner face.

Crico-arytenoideus dorsalis is an extensive sheet clothing the dorsum of the cricoid, separated from its fellow by the median crest on the cartilage.　Its fibres converge and proceed forwards and laterally, becoming concentrated towards their insertion on the muscular process of the arytenoid.

Crico-arytenoideus lateralis.　Fibres arise from the cranial edge of the cricoid arch and proceed dorsally and forwards to insert on the muscular process ventral to the preceding.　They form the more superficial lamina of fibres between the thyroid lamina and the conus elasticus.

Thyro-arytenoideus, a thin muscular sheet lying on a deeper plane than the preceding.　Springing from the deep surface of the thyroid lamina near its ventral border, fibres pass dorsally to insert on the lateral aspect of the arytenoid.　No specialized vocalis muscle was encountered, the vocal fold being purely membranous.

Arytenoideus is constituted by a thick mass of muscle occupying the dorsal concavity of each arytenoid, passing from one cartilage to the other across the interval between them.　No differentiation of oblique fibres was observed, all being transversely disposed.　No muscular fibres are carried into the ary-epiglottic folds either from the arytenoideus or thyro-arytenoideus.

2. TRACHEA (figs. 48, 49)

The tube measures 41 mm. long, i.e. the same as recorded for the largest male *Hapale* by Beattie, who found it somewhat shorter in females.　In diameter it measures 6 mm.　It is supported by twenty-five C-shaped cartilaginous rings.　The third and fourth rings are related to the thyroid gland.

The membranous dorsal wall is extensive, 2.8 mm. across at its widest part which is near the commencement, but narrowing to 1.8 mm. towards the bifurcation.

Bifurcation into right and left bronchi occurs in the upper thorax.　There is much asymmetry between the two bronchi, the left being longer and more oblique (15 mm.) and the right only 8.8 mm.

3. LUNGS

The left lung measures 38.7 mm. long, 20.5 mm. dorso-ventrally, and 13.5 mm. medio-laterally.　Corresponding measurements for the right organ are 37.5 mm., 27.0 mm. and 15.7 mm., so that in total bulk the latter is slightly in excess of its fellow.

The left lung is divided completely to its hilum into two (apical and basal) lobes.　The right is similarly divided into four lobes (apical, middle, basal, and azygos). Each lobe is provided with its own bronchus and vessels.

The left apical lobe presents a bluntly rounded apex, a smooth dorsal border and crenated ventral and caudal borders. The ventral border has an L-shaped indentation with a tonguelike lobe forming the caudal limb of the L. The fissure between apical and basal lobes is almost horizontal when the lung is standing in its base. The basal lobe is roughly cone-shaped, with an apex projected craniad dorsal to the entry of the lobar bronchus, around which the lung tissue forms a lappet on the face covered by the caudal margin of the apical lobe. The parietal border of the basal surface is somewhat crenated towards its dorsal end. The visceral and parietal pleura meet at the hilum and also along the ligamentum latum pulmonis, which is continued as far as the level of the diaphragm.

The right apical lobe is more quadrate in form than the left, presenting a double apex, the dorsal one being the true apex, which is separated from the secondary apex by the broad groove occupied by the subclavian vessels. The bronchus enters on the mediastinal surface about halfway along the dorsal border. The ventral border shows one deep crenation; the caudal border is obliquely disposed, only its dorsal moiety overlapping

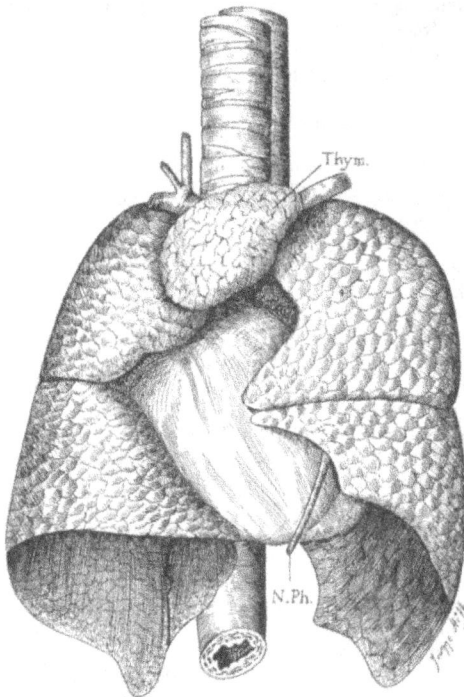

Fig. 49. Respiratory tract and structures in the mediastinum from the dorsal aspect.

the middle lobe superficially, leaving a triangular interval where the pericardium is exposed between the two lobes.

The middle lobe is wedge-shaped, with the apex antero-dorsally, where it is anchored by its own bronchus. The borders are anterior, postero-ventral and postero-dorsal, the last mentioned lying along and somewhat overlapping the cranial border of the basal lobe. Ventrally it contributes the ventral one-third of the diaphragmatic surface of the lung. The basal lobe is of pyramidal form, with its apex antero-dorsally, where, like the middle lobe, it is anchored by its own bronchus. Its base is formed by the concave diaphragmatic surface, of which it contributes the dorsal two-thirds (except for the small area supplied by the azygos lobe).

The azygos lobe is attached by a short bronchus to the ventro-mesial wall of the bronchus of the basal lobe. It is a small pyramidal lobe projecting from the medial aspect of the right lung caudal to the hilum and ventral to the ligamentum latum, which proceeds along the basal lobe immediately dorsal to the line of contact with the azygos lobe. The lobe presents medial, dorso-lateral, ventral, and basal surfaces. The medial surface forms part of the mediastinal aspect of the lung and is in contact with the pericardium. The dorso-lateral surface is applied to the basal lobe; the ventral aspect faces

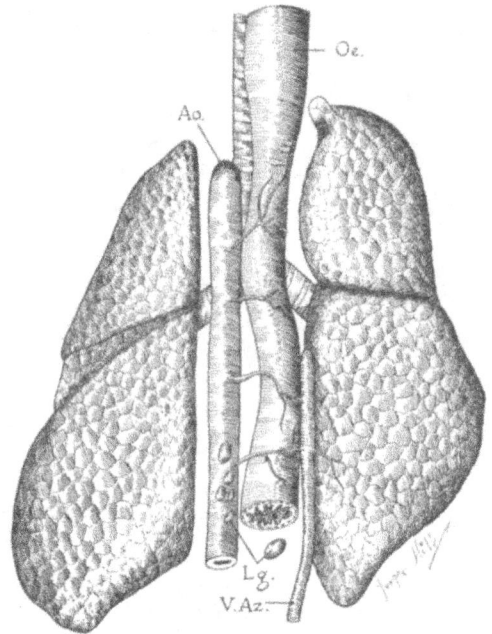

Fig. 48. Thoracic viscera from the ventral aspect.

the dorsal wall of the thoracic part of the postcaval vein and a rounded lobule projects from it around the left side of the vein, but no farther. The basal surface is flush with the diaphragmatic aspect of the basal lobe.

4. PLEURAE

Apart from the presence of the infracardiac bursa for the reception of the azygos lobe, the pleurae present little for remark. The cervical domes project a few millimeters anterior to the first rib. Posteriorly the left pleura gains the tenth rib in the mid-axillary line, the eleventh in the mid-scapular line and beyond the twelfth posteriorly. There is an invagination on the mediastinal side for the pericardium (see also under pericardium p. 82).

III. UROGENITAL SYSTEM (figs. 50–52)

1. KIDNEYS

As in *Hapale* the two kidneys lie virtually at the same level opposite the second and third lumbar vertebrae. Their measurements are given in table 7.

TABLE 7

	Left	Right
Sagittal length	20 mm.	20.7
Transverse	14.4	13.7
Dorso-ventral	7.9	9.0

The right is therefore somewhat the larger. Both are smooth surfaced, without trace of lobulation and covered with a fine capsule as well as a fascial sheath, the latter with its own vessels. In each the dorsal aspect is flat and the ventral strongly concave. In outline both have a broad cranial pole and a relatively acute caudal pole, the transition between the broad and narrow portions being at the junction of anterior third with posterior two-thirds of the gland. The hilum is located on the dorsal aspect, near the medial border and considerably nearer the caudal than the cranial pole; this is quite different from the arrangement in *Hapale*. In consequence there is no hilar notch along the medial border, which is therefore practically recti-linear. Structures enter the hilum in the usual relation to each other, but the obliquity of the renal arteries is remarkable (*vide infra*).

In section each kidney presents a cortex of 4.2 mm. thickness and a medulla of 4.5 mm., sharply demarcated. The renal tissue is divided into wedge-shaped masses by the distribution of the renal vessels, but the organ cannot thereon be described as multipyramidal, though clearly approaching that status. The collecting ducts, however, all open upon a single papilla, which is large but not projecting into the pelvis of the ureter.

2. URETERS

Within the renal sinus the ureter expands to form its pelvis which is not very large, but bifurcates into

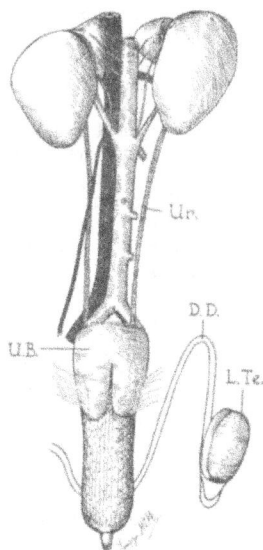

FIG. 50.   Urogenital system from the ventral aspect.

an anterior and a posterior subdivision. Outside the kidney the tube proceeds retroperitoneally along the dorsal abdominal wall, its length being 76 mm. As in *Hapale*, its course is in line with the apices of the lumbar transverse processes as far as the pelvic brim, but separated from them by psoas muscle. It crosses the common iliac vessels and has the usual relation to the ductus deferens before opening into the bladder.

3. URINARY BLADDER (figs. 50, 51, 52)

In the collapsed state this is an extremely elongated organ as in many other Platyrrhini. From fundus to neck the empty bladder measures 30 mm., all of which lies cranial to the pubes. The contracted organ measures transversely, just caudal to the fundus, 10 mm. The fundus is somewhat truncate rather than evenly rounded. The walls are extremely thick and muscular, especially towards the neck. The major part of the viscus has a serous covering, except near the neck and along a median ventral tract extending forwards to within 11 mm. of the fundus. Here the organ is provided with a short ventral mesentery or mesocyst as described by Förster (1922b) in several Cebidae. At the sides the serous covering extends 25 mm. towards the neck and dorsally 28 mm. There are well-marked lateral peritoneal folds. Dorsally a pronounced fold lies transversely immediately dorsal to the neck of the bladder, connecting together the terminal parts of the two ducti deferentes.

Internally the semi-opaque mucosa is fairly thick. Over much of the bladder it is thrown into pronounced longitudinal folds which converge towards the neck, but these are less evident dorsally towards the fundus than elsewhere. Here a papillated appearance is presented. The trigone, which occupies the dorsal wall immediately adjacent to the neck, is quite smooth. The trigone is a relatively small area, 4 mm. across its base (between the two rounded ureteric openings) and each lateral boundary 5 mm. long. There is no transverse ridge connecting the ureteric openings. The neck of the bladder is indicated by a constriction brought about by a sphincteric muscular thickening covered with a prominent mucous covering. There is no urachus.

4. URETHRA

From the neck of the bladder to the external urinary meatus the tube measures 44 mm. long, of which the first 20 mm. lies dorsal to the symphyseal region of the pelvis. The tube describes an S-shaped bend passing first directly backwards, bending ventrally beneath the subpelvic arch. Thence it turns forwards again, forming a hairpin bend, running deep to the scrotum and then some distance along the ventral surface of the symphysis before proceeding along the penis. The

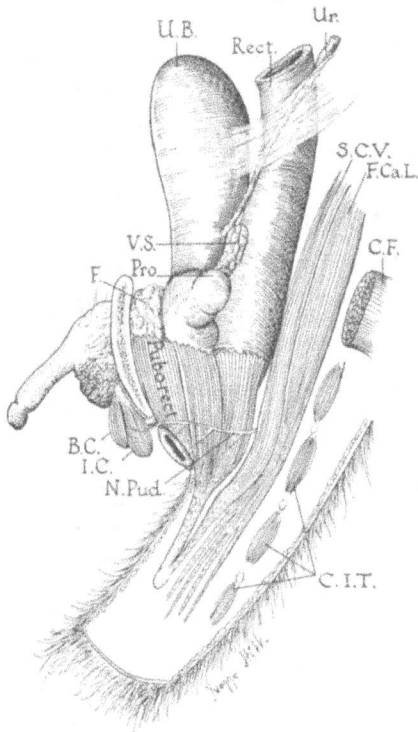

FIG. 51. Pelvic organs and related structures from the left side.

dorso-ventral gap between the caudally and cranially coursing segments of the tube measures some 15 mm.

The prostatic urethra accounts for the first 7–8 mm. of the tube. This lies against and is partly embedded in the ventral surface of the prostate. At its commencement it is roomy, but distally it narrows slightly. The dorsal wall projects into the lumen, forming a raised longitudinal fold immediately adjacent to the vesical sphincter, but narrowing rapidly thereafter. This elevation is all that remains of a colliculus seminalis. The chief peculiarity is its proximal position. A minute papilla, with openings of the ejaculatory ducts, lies on the most prominent part of the elevation, but no utriculus masculinus could be detected under the binocular microscope.

The following 10–12 mm. corresponds topographically to the human pars membranacea urethrae and is related to muscular fibres and fascial strata, but these do not constitute a uro-genital diaphragm in the sense understood in anthropotomy. Nevertheless, there is a well-defined fascial plane connecting the subpubic arch with the concavity of the urethral bend and, on the pelvic side of this, a narrow strand of muscular tissue passing over the crus penis to connect finally with the central tendinous point of the perineum.

Entering the bulb, the urethra thenceforth travels within the corpus cavernosum urethrae, lying in the topographically dorsal part thereof, sheltered in the median groove of the corpus cavernosum penis. It finally terminates in the glans in a small vertical slit-like meatus.

5. MALE REPRODUCTIVE SYSTEM

a. Testes

The testis, with the epididymis, occupies a peritoneal sac, the tunica vaginalis, to the dorsal wall of which it is connected. The cavity of the tunica does not ascend the spermatic cord. The testis and its coverings in the specimen examined were not completely scrotal in position, a condition found in many marmosets and tamarins, but over the caudal part of the inguinal region; but a fibrous cord connects the lower pole with the floor of the scrotum.

TABLE 8

	Cranio-caudal Length	Breadth	Dorso-ventral
L. Testis with epididymis	16.5	7.3	5.0
L. Testis alone	10.7	7.3	4.6

The measurements given in table 8 indicate the size (mm.) of the respective parts.

The testis is a compressed oval structure with the usual appearance. The dorsal border is covered by the corpus epididymidis, which is very slender. The corpus

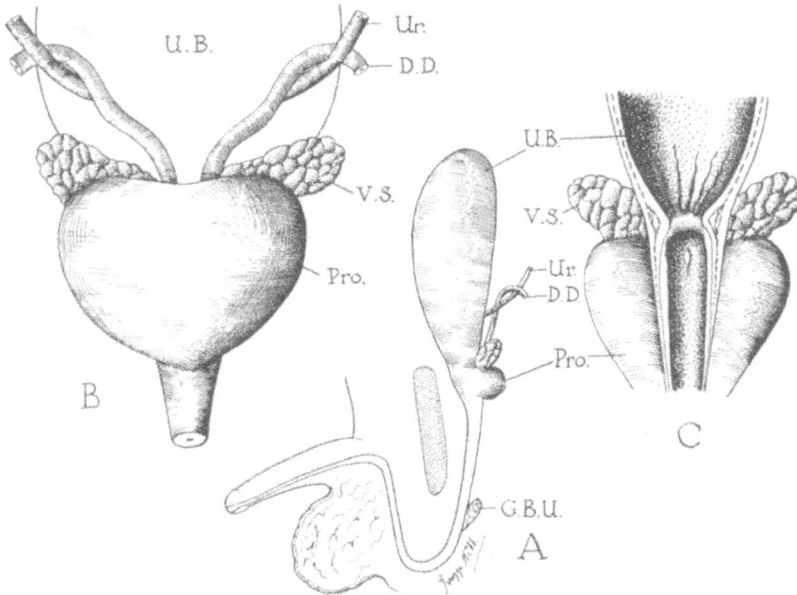

FIG. 52. Pelvic viscera. A. Diagrammatic representation of general arrangement of lower urinary tract.
B. Prostate and visiculae seminales from the dorsal aspect. C. Neck of bladder and prostatic urethra
opened from the ventral aspect.

enlarges at its two ends, overlapping the corresponding poles of the testis. The cauda epididymidis, much larger than the caput, is of triangular outline, much compressed laterally. The apex is adherent to the first part of the ductus deferens, and its base firmly adherent to the caudal pole of the testis. The caput is smaller and fairly free from the testis especially on the dorsal side, a fossa, lined by tunica vaginalis intervening. There is, however, no digital fossa between the testis and the body of the epididymis.

The ductus deferens commences in the cauda epididymidis and courses proximad along the corpus. It enters the spermatic cord which measures, from the upper pole of the testis to external abdominal ring, 8.4 mm. Another 13.7 mm. takes it to the internal ring, and from there to the base of the bladder the ductus measures a further 8.8 mm. Here it receives the duct of the corresponding seminal vesicle, the common ejaculatory duct so formed taking the usual course through the prostate to open into the urethra.

b. Seminal Vesicles

These are small lobulated bodies of brownish color situated dorsal to the neck of the bladder and in contact dorsally with the ventral wall of the rectum. They are much smaller than in *Tarsius* (Hill, 1955) and lie more transversely in the angle between the cranial margin of the prostate and caudal part of the bladder. Craniocaudally they measure only 5 mm. long.

c. Bulbo-urethral (Cowper's) Glands

This is a small paired lobulated structure of brownish color lying dorso-lateral to the urethra before the latter enters the bulb. Each measures 3 mm. long, but is compressed dorso-ventrally and firmly adherent to the urethral wall. Emerging from the distal end, the duct proceeds along the urethra to enter it just prior to the latter's entry into the bulb.

Microscopically the organ is a compound racemose gland with the groups of acini separated by a relatively large amount of connective tissue. In some of the larger connective tissue trabeculae a few striped muscle fibres occur, and others are present in the fibrous sheath of the gland, as described in prosimians by Oudemans (1892), but the gland lacks the saclike character found in *Lemur* and *Hapalemur* (Hill and Davies, 1954). The acini are lined by a single layer of low columnar epithelium.

6. FEMALE REPRODUCTIVE SYSTEM

a. Ovaries

As in other Platyrrhini these are large for the size of the animal, but in the specimen examined the right was considerably larger than the left (table 9).

Both occupy the usual position against the parietes overshadowed somewhat by the curving uterine tube. The long axis lies almost in the sagittal plane, with the fimbriae of the tube applied to the cranial pole. The caudal pole is connected to the tubo-uterine angle by a distinct ovarian ligament. Each ovary is connected

TABLE 9

	Right Ovary	Left Ovary
Length between poles	8.8	6.4
Breadth	4.5	3.5
Medio-lateral thickness	2.6	2.5

to the dorso-medial surface of the extensive broad ligament by a broad area of attachment. There is no definite mesovarium, the organ being sessile. Ovarian vessels proceed caudally near the free edge of the anterior portion of the broad ligament towards the cranial pole of the ovary, and these supply branches to the gland via the attached border of the organ. There is likewise no ovarian bursa, relation to peritoneum being as in Man (Zuckerkandl's type I).

In appearance when fresh each ovary is smooth, semi-translucent, and pinkish-brown in color, but the enlarged right organ was much browner than the left when fresh, especially at the cranial pole. At one point a hyperaemic locus was detected, viz. on the ventro-medial surface. This suggested recent ovulation, but there is no trace of a corpus luteum. In the fixed organ the parenchyma exhibits a uniform brownish tint throughout and is of soft consistence. The whole is covered by a firm tunica albuginea easily stripped off the parenchyma.

b. Epoophoron etc.

A soft whitish flattened body some 6.8 mm. diagonally lies in the mesometrium between attached border of the ovary and the lateral part of the uterine tube. It consists of about a dozen undulating tubules coursing at right angles to the long axis of the ovary. The tubules are of friable consistence. The more medial four or five tubules are somewhat distinct from the remainder and probably represent the paroophoron.

c. Genital Tract

Uterine tube

This does not course along the free edge of the broad ligament, but is enclosed between its layers some 5–6 mm. from the border. In consequence there is a narrow undulating membrane, the anterior mesosalpinx, cranial to the tube. This is continued medially to the angle between the tube and the fundus uteri; it is not continued as a membrane or even as a fringe on to the fundus as it does in some Cebidae (e.g. *Callicebus, Lagothrix*, Hill, 1952; 1959).

Within the mesosalpinx the tube takes a convoluted course. Measured from fundus to abdominal ostium

the tube covers a distance of 23.5 mm., but if unraveled would cover almost twice that distance. The wall of the tube is thick and firm. The abdominal end or infundibulum is, however, thin and delicate. It is applied closely to the cranial pole of the ovary and its fimbriae are few and short. The tube courses first away from the ovary and then, after a right-angled bend, turns mediad. This portion shows four or five close convolutions, only the first segment straightening as it approaches the uterus. The intramural portion is 3.3 mm. long.

Uterus

The organ exhibits the characteristic pyriform shape of the organ in higher Primates, having a highly convex globular fundus, a dorso-ventrally flattened body and a narrow, thick-walled cervix. Following are its measurements (table 10). Peritoneum covers the whole fundus and passes caudally on the ventral side about halfway along the length of the cervix. Dorsally the floor of the recto-uterine pouch reaches only to the level of the beginning of the cervix, i.e. the internal os.

TABLE 10

Length; fundus to external os	22.5 mm.
Maximum breadth	10.5 mm.
Minimum breadth (junction of cervix with body)	5.0 mm.
Maximum dorso-ventral diameter	6.6 mm.
Length of cervix	11.0 mm.

The shape of the uterine cavity conforms to that of its exterior. Dorsal and ventral walls are in contact and present two flattened areas of spongy highly vascular endometrium marked by the openings of the uterine glands. The texture of the endometrium of the corpus

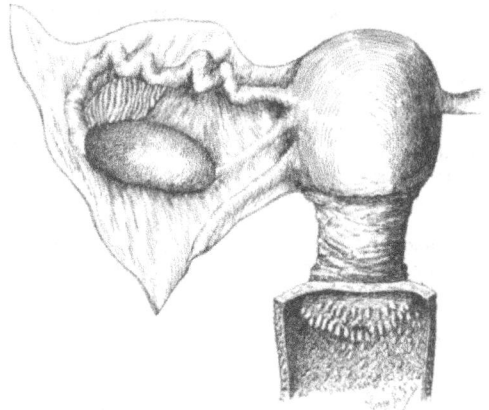

Fig. 53. *Callimico goeldii* ♀. Internal genital organs, viewed from the dorsal aspect. The dorsal vaginal wall has been opened to show the character of its lining and the papillated vaginal portion of the cervix uteri.

uteri is decidedly spongy. In the cervix it is firmer, paler and marked by longitudinal folds. The transition is abrupt at the level of the internal os.

The external os is shaped as in the human multiparous female—a transverse slit with rather ragged edges. The epithelium covering the vaginal portion of the cervix is papillated like that of the vagina, but the papillae are even more prominent, though less keratinized (fig. 53).

Vagina

This is a dorso-ventrally flattened passage, 30.0 mm. long, with a transverse diameter externally of 7.2 mm. It is compressed between the rectum dorsally and the pubic symphysis ventrally, for the urethra courses in the fibrous tissue of the ventral vaginal wall, external to the muscular coat.

Internally its mucosa is highly characteristic. It is strongly keratinized and marked by numerous areolae each topped by a horny papilla, each with a darkened apex. The areolae and papillae are arranged in more or less regular longitudinal columns along the long axis of the passage. Approximately 32 columns occur, about half on the dorsal and the remainder on the ventral wall. In the lateral sulci and in certain minor longitudinal sulci elsewhere there are gaps between the columns. The columns, on the whole, appear along raised longitudinal folds, but there is some slight interlacement between neighboring columns. Above the papillae tend to enlarge and become more conical; this tendency further increases on the vaginal portion of the cervix uteri. Dorsal and ventral fornices separate the vaginal wall from the corresponding cervical lips.

Vestibule

This forms the floor of the rima pudendi. It is covered with smooth, delicate rosy mucous membrane, presenting a number of slight longitudinal folds which converge towards the vaginal opening in the dorsal portion of the cavity. There is no abrupt transition from the vestibular towards the vaginal type of mucosa, the keratinized growths of the latter becoming evident gradually from without inwards. On the median line of the ventral part of the vestibule is a low longitudinal fold, with slighter lateral folds converging on to its sides. Dorsally, at the vestibulo-vaginal junction, is a more prominent conical elevation with its apex directed vertically. This is a urethral caruncle, though the urethra opens, not in its apex, but in the angle between the elevation and the vestibular wall, and therefore, hidden by the caruncle. From the caruncle two paramedian folds separated by a depressed area proceed into the vaginal opening.

IV. DUCTLESS GLANDS (figs. 41, 42)

1. SPLEEN

This is of normal size but peculiar shape, differing from that of all hapalid genera, being three-lobed, the lobes arranged in a T-formation with the cross-limb dorsally, and the long stem directed ventrad, embracing the fundus of the stomach. A sharp angle, with re-entrant notch, separates the long ventral limb from the antero-dorsal part of the cross-limb. The dorsal border, formed wholly by the cross-limb, is also slightly crenated. The posterior part of the cross-limb takes the form of a rounded bulge and is not demarcated so clearly from the dorsal part of the caudal border of the main stem, and its border shows no crenations. The two borders of the main lobe are smooth and parallel. They terminate ventrally by joining to form a bluntly rounded apex. The visceral surface has the usual relations to pancreas, etc., and to the peritoneum.

2. PITUITARY (fig. 63)

A flattened oval structure (26 mm. across) lying in the sella turcica between the two internal carotid arteries. Its largest part is the buccal 'lobe, which forms the whole of its dorsal contour, the neural lobe lying upon and partly recessed within the buccal lobe. The neural lobe is pear-shaped with the narrow end connected with the pituitary stalk. The gland receives paired branches from the internal carotids (fig. 63).

The gland was studied microscopically in sagittal sections cut at 10μ and stained with haematoxylin-eosin. The general arrangement of the different components of the organ agrees with that given by Hanström (1948, 1952, 1953) for the hapalid genera *Hapale* and *Leontocebus*, though differing in minor details from both (*vide* also Hill, 1957: 169, fig. 32). The gland shows the antero-posterior elongation and dorso-ventral compression observed in *Leontocebus*. In the sections the pars nervosa is rounded and presents the usual structure, with fusiform and multipolar neuroglia cells and fibres scattered in a homogeneous ground substance. Anteriorly local invasions by cells from the pars intermedia occur. Herring bodies were not observed.

The pars distalis (adenohypophysis) is relatively larger than in *Hapale* or *Leontocebus*, but does not cover the pars nervosa below to the same degree as in the latter. It consists of interlacing cell-columns separated by delicate connective tissue trabeculae carrying blood vessels (sinusoids). Chromophil and chromophobe cells are present in about equal numbers. Colloid aggregations are found in two locations—postero-inferiorly adjacent to the pars nervosa and at the extreme anterior end near the surface. Pars distalis extends beneath the pars nervosa to about the same extent as in *Hapale*, and therefore less than in *Leontocebus*, which Hanström states to be more like *Nycticebus*.

Pars intermedia is very distinct and separated from the pars distalis by remains of an intraglandular cleft,—not however forming Rathke's cysts as Hanström found in *Hapale*. It forms a thin layer between pars distalis and pars nervosa and spreads out on the dorsal aspect of the latter, agreeing in its proportions more with *Leontocebus* than *Hapale* as judged by Hanström's data.

Cells of the pars intermedia are very distinct, stain darkly and are closely packed. They have small cytoplasmic bodies with oval or rounded nuclei, but towards the intraglandular cleft the nuclei become very compressed and elongated. Some colloid bodies are present, especially above, and large blood vessels are evident. As already mentioned, locally the pars intermedia cells invade the substance of the pars nervosa.

Pars tuberalis extends upwards from the postero-superior part of the pars distalis along the pituitary stalk. Though continuous with pars distalis its structure is sharply demarcated. Its cells are large, cuboidal, and contain no cytoplasmic granules. Consequently, they are clearer in appearance than either type of cell met with in the pars distalis.

3. THYROID

This gland has the same general form and appearance as in Man, being pinkish-brown in color, lobulated, and very vascular. It consists of two lateral lobes 6.5 mm. long cranio-caudally, 2.8 mm. broad and 1.5 mm. thick. These lie along the wall of the most anterior part of the trachea and are connected across the ventral wall of that tube by a narrow, almost transparent isthmus some 4.5 mm. long. The isthmus lies opposite the interval between the third and fourth tracheal rings.

4. THYMUS

Some quite considerable remains of this exist in the anterior part of the thorax in the usual relation to the great veins etc. Of irregularly oval outline with marked asymmetry, the glandular tissue is composed of small polyhedral lobules held loosely together by connective tissue. The organ fails to reach the pericardium.

5. ADRENALS

Both lie in contact with the medial border of the cranial pole of the corresponding kidney. The left has a hollow base applied to the kidney and a rounded periphery. The right differs in its triangular outline but its base has a more rounded outline. Their measurements (mm.) are given in table 11.

TABLE 11

	Left	Right
Cranio-caudal l.	5.2	6.3
Transverse diam. at base	10	9.8
Dorso-ventral thickness	4.3	6.5

Vessels derived from the aorta and from the renal arteries course over them in the surrounding areolar tissue.

In section cortical tissue makes up the bulk of each gland, being 1.5 mm. thick. The section is triangular rather than triradiate in outline, but the medulla has a triradiate appearance within the broad cortex. The medulla is greyish in color and the cortex pinkish-yellow.

VASCULAR SYSTEM

I. PERICARDIUM AND HEART (figs. 48, 54 A, B)

1. PERICARDIUM

The pericardium is clothed externally on all sides by parietal pleura, except where the pulmonary vessels emerge from it to enter the lung on each side, and along a median dorsal strip where the sac abuts on the dorsal mediastinum. Ventrally the pleural membranes of the two sides meet along a line running antero-posteriorly along the pericardium, where they join to form a membranous mediastinal septum connecting the pericardium with the sternum. Caudally these layers are continued as a strand connecting the pericardium over the cardiac apex to the diaphragm. The only other connection with the diaphragm is by two short two-layered antero-posterior septa which form the lateral boundaries of the so-called infra-cardiac bursa of the right pleura, which lodges the azygos lobe of the corresponding lung. The left phrenic nerve raises a fold of pleura as it courses over the left wall of the pericardium. The right nerve lies much more dorsally.

There is a well-marked transverse sinus in the pericardial cavity, but the oblique sinus is absent on account of the primitive arrangement of the pulmonary veins (p. 84). Its site of development is marked by a slight dimple on the posterior wall of the serous pericardium in the angle to the left of the termination of the postcaval vein and the great venous bay formed by the union of the pulmonary venous trunks.

Over the aorta there is much subepicardial tissue, and more of this, with some fat, occurs in the coronary sulci and along the main stems of the coronary vessels. The pulmonary artery, on the contrary, is free from this development. Pericardium extends to include the proximal 3.2 mm. of the great vessels.

TABLE 12

Maximum length from base to apex	20	mm.
Maximum breadth across ventricles	14.5	
Total height from diaphragmatic surface	15.0	
Length of margo acutus	16.5	
Thickness of right ventricular wall	2.5	
Thickness of left ventricular wall	4.0	

The heart lies more sagittally in the thorax than in Man. In this it agrees with those of the prosimians (Hill & Davies, 1956). Nevertheless, the apex is directed caudally and to the left, but it is very rounded, as also are both margo acutus and margo obtusus, though both these are recognizable. The principal dimensions are given in table 12.

2. DESCRIPTION OF THE CARDIAC EXTERIOR

The ventral aspect is comprised chiefly by the right ventricle, recognized from the left by its darker color.

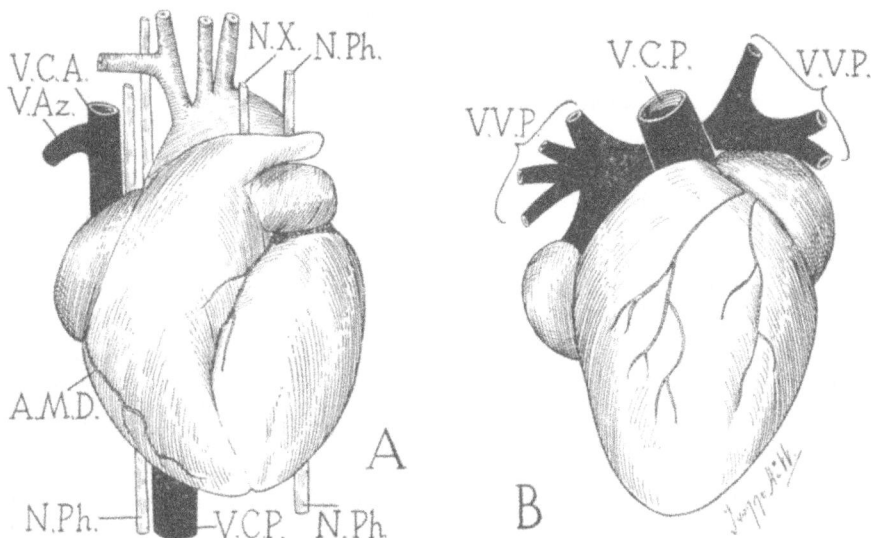

FIG. 54.   Heart and great vessels.   A. From the ventral aspect.   B. From the diaphragmatic aspect.

There is no interventricular sulcus, but the site of it is marked by vessels and a line of fat deposit. The ventricle narrows cranially to become the infundibulum which gains the anterior part of the left margin. Remainder of the left margin, including the whole apex, is formed by left ventricle.

The ventral end of the right auricular appendage forms a triangular area to the right of the base of the right ventricle, notched just below its middle overshadowing the coronary sulcus.

A still smaller amount of the left auricular appendage appears on the ventral aspect to the left of the infundibulum, where it deeply overshadows the coronary sulcus.

The dorsal aspect is comprised largely by left ventricle, with a small part contributed by the right ventricle near the margo acutus. Towards the base the two atrial chambers form a narrow strip, continued laterally by the pulmonary venous trunks radiating from the median common sinus. There is a semilunar area to the right pertaining to right atrium, with the postcava terminating in this at the caudal extremity (fig. 54, B).

The base is formed by the two atrial chambers, especially the left, which, though small in capacity, is expanded by the formation of a common venous bay or sinus formed by the union of all the pulmonary veins from the two sides. To the right lies the smooth-walled dorsal portion of the right atrium receiving the precaval and postcaval veins in front and behind respectively. A distinct shallow sulcus termi-

nalis connects the right margins of these two orifices and serves to separate the atrium proper from its relatively voluminous auricular appendage.

3. INTERIOR OF THE CARDIAC CHAMBERS

The right atrium presents the usual two parts, a smooth-walled sinus venosus dorsally and an auricular appendage ventro-medially. The two are separated by a very faint sulcus terminalis on the outside, but more definitely internally by a ridge (crista terminalis). The appendage is a small pocket with trabeculated interior, owing to the musculi pectinati radiating from the crista. The sinus is lined by smooth endocardium which, at the opening of the posterior vena cava, is thrown into two large semilunar folds which constitute a valve. One fold situated dorso-medially ends on the interatrial septum in an elongated cornu dorsal to the opening of the coronary sinus. The right or lateral fold is placed more ventrad and its medial horn proceeds to the ventral margin of the coronary sinus opening, and like its fellow is prolonged as a pillar-like thickening beyond the opening. Between the two, guarding the posterior edge of the oval coronary sinus opening, is a small cross-fold (Thebesian valve). This arrangement is unique among the Primates so far examined in this connection as in none of the Strepsirhini examined by Davies (1947) or Davies and Hill (1956) was this observed. The nearest perhaps are *Propithecus* and *Hapalemur* where the two valve cusps embrace medially the coronary sinus opening, whereas in the Lorisoidea usually the cusp or cusps, where present, terminate dorsal to that opening. Moreover in *Callimico* the semilunar folds are on the whole more obliquely placed than in the prosimians instead of being directed more dorso-ventrally.

Even in the Hapalidae no exact counterpart of the arrangements has been seen, though it must be admitted there is a fair degree of individual variation in a single species. Thus in two specimens of *Hapale jacchus* one had a single semilunar valve cusp on the dextral side, with its medial (ventral) horn passing beyond the opening of the coronary sinus, over which it formed a partial valve, to end on the ventral limbus of the fossa ovalis. In another specimen a well-developed septal fold was also present, its attached border aligned along a fleshy pillar which skirted the dorsal margin of the coronary sinus; in this case the dextral fold terminated on the ventral margin of the sinus opening; no intermediate valve like that in *Callimico* was present. In an example of *H. penicillata* the arrangement was as in the first-mentioned *H. jacchus*, but the fold was located farther from the coronary opening. To the dorsal side of the coronary opening was a fleshy pillar, but this had no septal valve associated with it, except perhaps as a vestige represented by a concave elevation running towards the dorsal wall of the atrium from the caudal end of the fleshy pillar.

The nearest approach to the arrangement in *Callimico* was found in an *Oedipomidas oedipus* where septal and lateral crescentic folds are present, their medial horns related to the dorsal and ventral borders of the coronary sinus opening respectively. But the dextral fold is much more obliquely disposed so as to overshadow the ventral half of the sinus opening, its free border passing almost sagittally. There is no intermediate cuspule guarding the caudal end of the coronary sinus.

Tamarinus mystax, Tamarin midas, and *Leontocebus rosalia* agree in the possession of a very extensive, but sometimes fenestrated, always thin septal cusp. The dorsal horn of this fold arises far forward on the atrial wall, the free edge sweeping postero-medially to attach to the septum *ventral* to the coronary sinus opening, crossing a fleshy pillar which guards the dorsal edge of that orifice. *T. midas* also shows an equally extensive dextral cusp ending with the septal cusp on the ventral pillar of the limbus fossae ovalis, but the two others lack the cusp entirely.

Other details of the interior of the right atrium of *Callimico* include the vague outline of the fossa ovalis, the limbus being discernible only cranially, where it carries a vessel, and caudally. Anterior to the cranial part of the limbus is a well-marked crescentic fold caused by the projection inwards of the atrial wall in the angle between the ventral wall of the anterior caval vein and the cranial wall of the auricular appendage. This fold forms a partial septum and in its free border carries a fairly large artery coursing from the dorsal atrial wall towards the crista terminalis.

The *left atrium* is lined by smooth endocardium throughout. The walls are muscular, especially in the appendage, where the muscular layer abruptly thickens, but without forming musculi pectinati. The fenestra ovalis is more sharply demarcated than on the right. The chief peculiarity of the left atrium is the dorsal vestibular, saclike extension which forms a common bay for the reception of the pulmonary veins from both lungs.

The *right ventricle* is a pyramidal chamber with a relatively smooth, convex septal wall and trabeculated dorsal and ventral walls. Two papillary muscles are present, a short one attached rather near the base of the ventricle at the junction of the septal and ventral walls, the other longer, situated towards the apex to the right side of the ventral wall. These between them supply chordae tendineae to the right and left cusps of the tricuspid valve. The broad septal cusp is connected to the septal musculature directly by numerous short chordae. The right papillary muscle receives on the left side of its base a thick columna carnea from the septal wall, serving as a substitute for a moderator band and doubtless carrying conducting tissue.

A crista supraventricularis is present, arching over the base of the ventricle ventral to the opening of the infundibulum.

The cusps of the pulmonary valve are thick and robust, but without corpora Arantii. They are arranged one dorsally and to the left; one ventrally and to the left and one on the right and slightly ventrad.

The *left ventricle* is provided with relatively few but short, thick columnae carneae. Papillary muscles are not differentiated, the chordae tendineae arising from longitudinally disposed columnae on dorsal and ventral walls of the chamber.

The cusps of the aortic value are arranged one dorsally and two ventrally, left and right. They are provided with corpora Arantii.

II. PULMONARY CIRCULATION

From the infundibulum the pulmonary artery proceeds craniad and to the left in close relation to the first part of the aorta. A short fibrous connection (the remains of the ductus arteriosus) joins the pulmonary trunk to the ventral wall of the aorta just craniad of the origin of the left coronary artery. The trunk bifurcates in the concavity of the aortic arch into right and left pulmonary arteries. At the hilum of each lung the arteries break up into lobar branches. In each case the veins lie mainly ventral to the arteries and the bronchi dorsal to them, but the apical branch of the left artery proceeds cranial to the corresponding vein, arching over the left bronchus to enter the apical lobe; at the same time the artery to the basal lobe proceeds dorsal to the corresponding bronchus, so that the relations at the hilus of the basal lobe are dorso-ventrally artery, bronchus, vein. On the right side all the veins are ventrally placed, the arteries next, then the bronchi;

but the artery and bronchus to the basal lobe are again reversed, the artery entering dorsal to the bronchus. The artery to the azygos lobe lies medial to the corresponding bronchus.

Beattie remarks upon peculiarities in the arrangement of the lobular veins in *Hapale;* those from the right lung joining to form a single trunk which opens on to the dorsal wall of the left atrium, while on the left side the two lobular veins open separately although closely approximated. He regards this as an advance on the arrangement in *Tarsius. Callimico* retains the tarsioid condition, where all the lobular veins discharge into a common chamber or vestibule on the dorsal aspect of the left atrium. Three lobular veins open into the right side of the vestibule, draining apical, middle, and basal (with azygos) lobes respectively. Four veins discharge into the left side of the vestibule, one from the apical and three from the basal lobe. The condition is paralleled in some prosimians (Hill and Davies, 1956).

III. SYSTEMIC ARTERIES

1. AORTIC ARCH

The aortic arch (fig. 54, A.) is of the broad, low type, the span measuring 11 mm. and the height of the arch 4.7 mm. (measured according to specification given by Hill and Davies, 1956, in discussing the prosimian aorta). The arch has the usual relations. The ascending portion presents well-marked sinuses of Valsalva arranged opposite the corresponding valve-cusps. From the left and right ventral sinuses spring the corresponding coronary arteries.

The transverse portion of the arch gives rise from its summit to three branches, innominate, left common carotid, and left subclavian. The two former arise in contact with each other, virtually from a common root, while the left subclavian is separated from the left side of the left common carotid by a distance of 0.5 mm. The innominate is, moreover, very short, breaking up almost at once into right subclavian and right common carotid. The arrangement accords most nearly to Keith's (1895) type C, which approaches the condition common in Strepsirhini such as *Perodicticus, Arctocebus, Galago, Lemur,* and *Hapalemur* where only two vessels spring directly from the arch (Hill and Davies, 1956). *Hapale,* according to Beattie (1927), has three branches from the arch, but no details are given as to their relative positions. Among Cebidae, on the other hand, Keith (*loc. cit.*) refers to three examples of *Alouatta* and six of *Ateles.* Of the former, two showed branching of type B (only two branches from the arch) while the third pertained to type E (three branches widely spaced, the normal human arrangement). In *Ateles,* however, Keith found three with type A arches (two branched with the left subclavian arising very near the innominate) and three with type C. There is evidently much variation among the

Cebidae, a conclusion which is supported by observations on two specimens of *Cebus.* In an adult female *C. apella* (Brown capuchin) an arch of intermediate type was found—three branches all bunched together on the summit of the arch. In another, a hybrid between a female Brown capuchin and a Smooth-headed capuchin (*C. apella xanthosternos* ♂), an arch of type D was present.

My own investigations in the different hapalid genera may be summarized thus:

1. Hapale

In three specimens of *Hapale jacchus* the aorta virtually gives rise to two vessels only, for the left common carotid arises from the left side of the innominate; the left subclavian, moreover, springs from the angle between the origin of the innominate and the summit of the arch. Beattie's description, therefore, does not constantly apply in this species, all three examples here mentioned being referable to Keith's type A.

A single female of *H. penicillata* falls into category B—a two-branched aorta with the left subclavian arising 1.4 mm. to the left of the innominate. The left common carotid arises from the innominate 2.0 mm. from its origin, the innominate, thereafter, bifurcating after another 2.9 mm.

2. Mico

In a male of *M. argentatus* an almost identical pattern was found to the *H. penicillata* just described, but the first part of the innominate measures 2.5 mm. and the second 3.5 mm.

3. Cebuella

In a Pygmy marmoset (*Cebuella pygmaea*) a three-branched arch was found, the three well-spaced (type E) as in Man. The innominate bifurcated after 4.5 mm.

4. Oedipomidas

In *Oedipomidas oedipus* the type mentioned in *H. penicillata* is preserved, but the left subclavian emerges from the angle between the arch and the left wall of the left common carotid, which itself springs from the angle between arch and innominate.

5. Leontocebus

In a male of *L. rosalia* a similar arch to that of *Oedipomidas* was found, but the left common carotid was more obliquely disposed, giving a more acute angle between itself and the summit of the arch; from this angle the still more oblique subclavian took origin.

6. Tamarin

Two examples of *Tamarin midas* examined fall into two different categories. In a female the pattern is almost identical with that just described for *Leontocebus rosalia* (intermediate between types B and C), the other, a male, falls between types A and B with the left common carotid springing from the left side of the innominate near its root, but more distinct than in *Hapale jacchus*. The left subclavian emerges here from the angle between carotid and summit of arch. The innominate measures 9.7 mm. long in the female with the type B–C arch, but only 4.4 mm. in the male with the A–B arch.

7. Tamarinus

In a female *T. mystax* three arteries spring from the arch, but these are not spaced apart as in *Cebuella*. The innominate, 4.2 mm. long, is succeeded immediately by the left common carotid and this again by the left subclavian, an arrangement which is intermediate between Keith's types C and D.

2. CORONARY ARTERIES

These arise in the usual way. The left breaks up almost at once into ventral (interventricular) and dorsal (circumflex) branches. The former again immediately trifurcates, giving off two large interventricular branches which proceed towards the apex and a smaller infundibular branch to the wall of the infundibulum. The dorsal branch, after supplying an atrial artery to the ventral wall of the left atrium, eventually gaining that of the right atrium, proceeds deeply in the coronary sulcus, hidden by the overhanging left atrial appendage. Few of the arteries run for long in the subepicardial tissues, but proceed for the most part intramuscularly.

The right coronary is a circumflex vessel occupying the right coronary groove. It gives off a number of infundibular and other branches to the ventral wall of the right ventricle, and also a large right marginal artery. It terminates as the dorsal interventricular artery.

3. DORSAL AORTA AND ITS BRANCHES (figs. 50, 55, 57)

The dorsal thoracic aorta measures 27.5 mm. long and the abdominal portion 50 mm., but the division between thoracic and abdominal parts is very indefinite on account of the great length and obliquity of the pillars of the diaphragm, which clothe the vessel laterally over a distance of 23.7 mm. The above measurements are taken from the point of first emergence from the pillars, i.e., just prior to the origin of the coeliac axis. Just beyond the arch the vessel has a diameter of 3.5 mm.; immediately prior to the bifurcation the diameter is reduced to 2.3 mm. The bifurcation takes place opposite the disc between the bodies of the last two lumbar vertebrae. The angle formed by the diverging common iliacs at the bifurcation is 50°.

From the descending thoracic aorta are given off seven pairs of aortic intercostals, supplying the posterior seven interspaces, and two mediastinal branches. The intercostals do not arise symmetrically, the left members of each pair springing slightly craniad of their

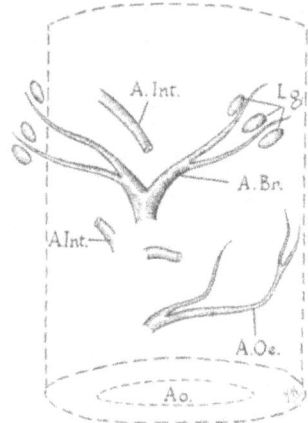

FIG. 55. Portion of thoracic descending aorta from the dorsal aspect.

fellows, but the asymmetry becomes progressively less in the hinder pairs, and has almost disappeared in the case of the most caudal pair.

In a typical intercostal space (fig. 71) the artery lies between the vein and the nerve, the vein lying nearest the subcostal groove. Each artery, from a point near where it is crossed by the sympathetic chain, gives a branch to the succeeding costo-vertebral joint and soon after a long, slender twig which proceeds obliquely towards the rib bordering the space caudally. It thenceforth changes direction and runs along the pleural surface of the rib, presumably to anastomose ventrally with a corresponding twig from the internal mammary (or musculo-phrenic in the case of the hinder spaces). The mediastinal branches proceed at first dorsally, but divide early, their tributaries being distributed to the oesophagus and neighboring connective tissues. In addition the anterior member divides symmetrically into two main branches, each of which further bifurcates almost immediately, the resulting vessels becoming bronchial arteries, which proceed along the bronchi to enter the corresponding lung hilum. Hilar lymphatic nodes are closely adherent to these vessels.

In the abdomen the descending aorta gives off firstly a pair of subcostal arteries and then five pairs of segmental lumbar arteries—all with the usual arrangement and distribution. A longitudinal anastomotic chain is formed between the segmental vessels along the lateral border of the ilio-costalis muscle deep to the internal oblique and superficial to the transversalis abdominis. The subcostal turns caudally to contribute a large share to this longitudinal channel.

Of the median visceral branches the coeliac and anterior mesenteric spring by a common stem soon after the aorta emerges from between the crura; the posterior mesenteric arises some 25 mm. caudally. Other visceral branches include the inferior phrenic, which is essentially adrenal in distribution, and paired middle adrenals and renals.

Coeliac Axis (figs. 56, 57)

This breaks up almost immediately into six branches; these are, from left to right in a clockwise manner: splenic, two left gastrics, hepatic, pyloric, and anterior mesenteric (see fig. 56). The splenic courses leftwards along the dorsum of the pancreas. It divides almost at once into two parallel branches, (a) a larger anterior (splenic proper) which takes a sinuous course along the cranial edge of the pancreas, terminating in the spleen. It gives anastomotic branches to the more caudal companion (b) which is purely pancreatic in distribution (pancreatica propria). In addition to pancreatic and splenic distribution the former vessel gives a large gastric branch which proceeds *via* the gastro-splenic omentum to the dorsal aspect of the fundus, dividing into two terminal branches.

Of the two left gastrics the sinistral one proceeds towards the left of the cardia, supplying the region of the cardio-fundic notch and the cranial wall of the fundal sac, extending transversely at right angles to the terminal gastric twigs from the splenic artery. The dextral left gastric makes for the right side of the cardia, to which it supplies twigs, thereafter proceeding between the layers of the lesser omentum, where it forms an anastomotic arcade with a similar vessel from the hepatic artery.

The hepatic artery is a large vessel proceeding to the lesser omentum directly but giving twigs to the lesser curvature in passing. It divides into right and left branches at the porta hepatis. The pyloric artery is distinct from the preceding and supplies the region of the pylorus and cranial part of the duodenum, also

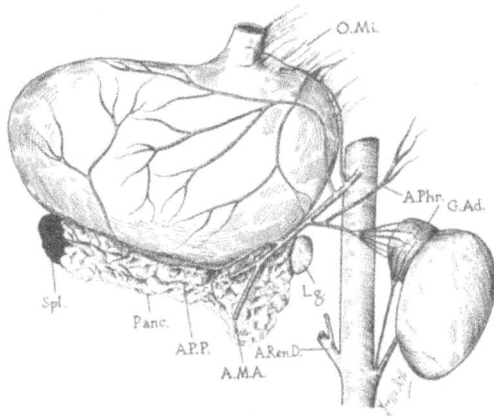

FIG. 57. Anterior part of the abdominal aorta and its branches from the ventral aspect. The stomach and pancreas have been reflected towards the right.

the caudal part of the bile duct and neighboring parts of the head of the pancreas.

The greater curvature of the stomach is supplied by vessels from a gastro-epiploic arcade derived from the splenic on the left and the gastro-duodenal on the right. The latter springs from the pyloric artery on the dorsal aspect of the pyloric region of the stomach and, after giving the right gastro-epiploic, continues along the left border of the duodenum (as the anterior pancreatico-duodenal artery) to anastomose with the most cranial branch of the anterior mesenteric. The arcade so formed gives a veritable plexus of arteries to the duodenum and neighboring part of the pancreas and common bile duct. It also gives some substantial communications to the root of the middle colic artery; these course in the connective tissue between the serous adhesions connecting mesocolon and mesoduodenum.

Anterior Mesenteric Artery (figs. 42, 56, 57)

This is a large branch of the coeliac axis and serves the intestine from the middle of the duodenum to the proximal part of the colon. After giving off the posterior pancreatico-duodenal, it supplies about a dozen *rami intestini tenuis*. These form a single series of arcades situated near the mesenteric border of the gut. At one point only, near the aboral end of the ileum, is an attempt made at a second arcade distal to the main series. The terminal intestinal twigs proceed from the arcades and bifurcate to send branches on each side of the intestine.

The anterior mesenteric terminates by dividing into two ileo-colic vessels; these run almost parallel, diverging but little, and then each divides into two in a plane opposite to that of the first division. The products of the second division proceed respectively on to the

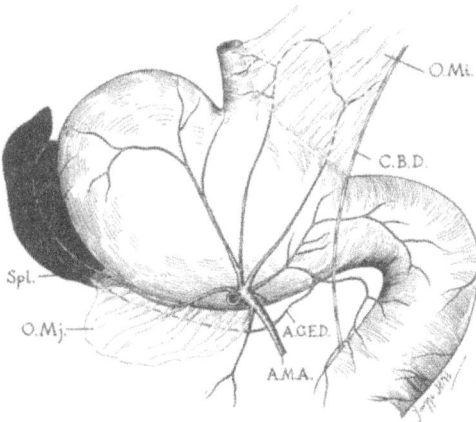

FIG. 56. The coeliac axis and its branches as viewed from the dorsal side.

dorsal and ventral aspects of the gut over the ileo-colic junction where they form two loops by joining together. From these dorsal (dextral) and ventral (sinistral) loops symmetrical branches are supplied to ileum, caecum, and proximal colon. The aboral of the two original vessels also gives off from its point of primary bifurcation an "ascending" colic branch, which forms a communication with the mid-colic artery. Caecal vessels show a primitive simplicity, being perfectly symmetrical and of equal size on dorsal and ventral sides. Each caecal artery courses in a slightly raised sero-fatty fold, but the dorsal fold is slightly tethered in the ileo-colic angle by a short, thick serous fold which evidently represents the mesotyphlon. About five to six terminal branches are supplied to the proximal colon by the more caudal of the two ileo-colics. The terminal ileum is supplied from the ileo-colic loop. There are no recurrent ileal vessels from the caecal arteries.

The *mid-colic artery* arises from the very commencement of the anterior mesenteric if not directly from the coeliac axis. It is short, dividing early into right and left primary branches. These diverge sharply in the root of the mesocolon where the latter is shortest. The right branch proceeds along the mesocolon near its attachment to the transverse colon, following the gut; it anastomoses with the above-mentioned ascending colic branch from the aboral ileo-colic. The arcade so formed gives numerous branches to the colon. The left branch similarly follows the arch of the colon on the left, finally uniting with the proximal colic branch of the posterior mesenteric.

Posterior Mesenteric Artery

This springs from the aorta some 25 mm. posterior to the coeliac axis. It enters the mesentery of the hind gut, giving off a proximal colic branch which runs transversely, but which shortly divides into ascending and descending branches. The former unites with the left branch of the mid-colic as already described. The latter proceeds caudad, forming a loop with the next main branch of the posterior mesenteric. At least three other colic branches arise more posteriorly, each progressively shorter than the last. No further loops are formed in the mesentery.

Adrenal and Renal Arteries (figs. 50, 57)

The adrenal is extremely richly provided with arteries. A pair direct from the aorta proceeds respectively to the apical region of the gland. On the left the vessel is continued forwards along the left crus as a posterior phrenic artery. This is a very long and attenuated vessel which curves to the right on gaining the middle arcuate ligament. As in *Tarsius* (Hill, 1953) there is no corresponding right phrenic. Following the above a series of smaller, shorter branches, with wavy courses, proceed to the middle portion of the corresponding adrenal. Finally, a pair of very long, thin vessels

springs from the roots of the corresponding renal arteries, proceeding forwards and somewhat laterad. They supply the hinder portions of the adrenal on both dorsal and ventral aspects and form a plexus on the concave renal aspect of the organ.

Renal arteries are peculiar in their recurrent course. Springing from the sides of the aorta about midway between the coeliac axis and the posterior mesenteric, both at approximately the same level, they turn forwards and laterad at an angle of 20° to the sagittal axis. From their anterior walls spring the posterior adrenal vessels, that on the left from the angle between the renal and the aorta, but that on the right somewhat more distally. From their posterior walls spring the spermatic arteries, which accompany the ureters to the pelvis and then skirt the side wall of the pelvis to gain the inguinal canal, whence they proceed to the corresponding testis. The main renal does not divide further until after it enters the hilum of the kidney, where it breaks up into interlobar vessels.

4. ARTERIES OF THE HEAD AND NECK
(figs. 58, 59, 60, 61)

Common carotid vessels arise in the manner indicated on page 85. They proceed forwards in the usual relation to neighboring structures, within the carotid sheath, as far as the angle of the mandible, deep to which the main artery is continued to the tympanic bulla as the internal carotid. At the angle of the mandible a number of branches is given off, all more or less

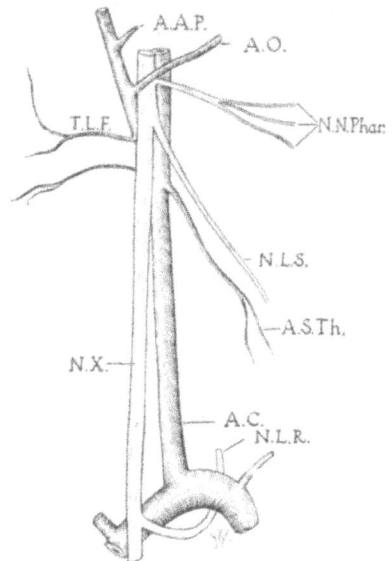

FIG. 58.   Left common carotid artery and vagus nerve.

at once, namely superior thyroid, occipital, and a large linguo-facial trunk. The external carotid is the cranial continuation through the parotid gland. While within the gland the internal maxillary artery is given off anteriorly. At the superior limit of the gland, opposite the neck of the mandible, the carotid bifurcates into a large posterior auricular and a smaller superficial temporal, the latter giving off anteriorly, almost immediately, a transverse facial artery.

The *superior thyroid artery* is given off dorsally. It takes an archlike course on the inferior constrictor, towards the cranial pole of the lateral lobe of the thyroid gland. Here it divides, giving branches to the ventral part of the gland, but continuing along its dorsal border, between the gland and the oesophagus, supplying both. It also contributes the superior (anterior) laryngeal artery (p. 74).

The *linguo-facial stem* courses along the cranial border of the intermediate tendon of the digastric muscle. It supplies, from the earlier part of its course, branches to the submandibular salivary gland and a submandibular branch along the lower border of the mandible. It also gives off vertically a branch which runs upwards deep to the internal pterygoid muscle. This is the ascending palatine which supplies branches along the palato-pharyngeus to the pharyngeal wall and above this directly to the superior constrictor and tonsil. It finally ends in the vault of the pharynx by anastomosing with a vessel running from before backwards from the nasal fossa, but whose exact source was not determinable in the material available, though probably representing the Vidian artery (a. canalis pterygoidei) of human anatomy.

The parent vessel, near the anterior border of the masseter, divides finally into facial and lingual trunks. The former could not be followed to its final distribution on account of the removal of the facial tissues in skinning. The lingual takes the usual course into the tongue after giving off an ascending branch towards the external pterygoid muscle, which it reaches along the lingual nerve. This vessel anastomoses with a branch from the internal maxillary.

The *internal maxillary* leaves the parotid gland on its deep surface and courses over the lateral surface of the internal pterygoid muscle. The only branches determined with certainty were the large inferior dental, accompanying the nerve, a branch accompanying the lingual nerve, mylo-hyoid twigs, twigs to the pterygoid muscles and the temporalis and the buccal artery, accompanying the long buccal nerve. The artery ends by becoming the infraorbital artery which follows the maxillary division of the Vth nerve on to the face, supplying the superior dental vessels as it proceeds. The above-mentioned Vidian artery presumably originates from the infraorbital continuation of the internal maxillary, as do also the vessels on the nasal septum and lateral nasal walls. There is also a large descending palatine artery.

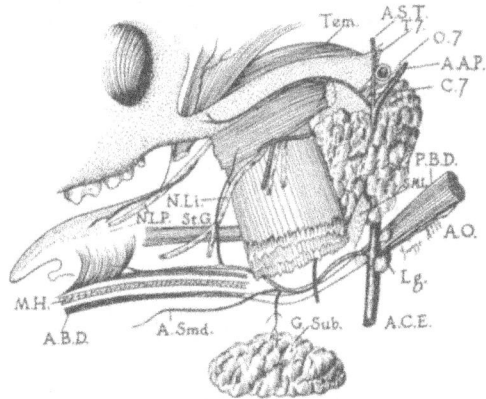

Fig. 59. Dissection of the left pterygoid region and neighborhood showing salivary glands and muscles of mastication, with neighboring arteries and nerves.

No meningeal branches were found and none were found injected outside the dura, though grooves are present on the inner surfaces of the cranial bones. Branches of the middle cerebral are visible through the dura and could easily be confused with meningeal arteries.

The only sizable meningeal vessel that could be determined with certainty was derived from the lachrymal which takes a recurrent course, entering the anterior fossa of the cranial cavity through a small foramen situated at the lateral extremity of the orbital plate of the frontal bone, where the bone unites with the lesser wing of the sphenoid. On entering the cranial cavity it divides at once into anterior and posterior branches. The former proceeds forwards in the groove between the orbital plate and the side wall of the brain case. The posterior larger branch evidently assumes the role played in Man by the middle meningeal. This agrees with the findings of Theile (1852) in *Macaca sylvana* and in *Mandrillus*, of Rojecki (1889) in two species of *Macaca* and of Tandler (1899) in *Hapale penicillata* (*fide* Mensa, 1913).

In addition to the preceding a small posterior meningeal artery was found in the anterior compartment of the jugular foramen between the ninth and tenth cranial nerves. This presumably corresponds to the vessel which, in Man, is derived from the ascending pharyngeal. Unfortunately, its connections had been severed before the contents of the jugular foramen were dissected.

Occipital artery: This is a derivative of the posterior aspect of the carotid system. It proceeds dorsally along the lower edge of the posterior belly of the digastric muscle, eventually appearing in the suboccipital space deep to the sterno-mastoid and trachelo-mastoid. It sends branches caudally along the deep surface of the complexus, one to the suboccipital triangle, which it gains by crossing the groove between the insertion of the inferior oblique and origin of superior oblique, and finally an ascending branch which runs along the lateral border of the superior oblique (which it supplies) to become superficial on the scalp.

Transverse facial artery: A slender vessel which courses along the zygomatic arch, anastomosing anteriorly with branches from the facial.

Posterior auricular artery: A very large vessel which is accompanied by the occipital branch of the seventh nerve.

Superficial temporal artery: Much smaller than the preceding. It terminates in the skin and subcutaneous tissues of the scalp anterior to the ear.

Internal carotid artery: The main artery gains the tympanum by passing through the large foramen on the medial wall of the bulla as in *Tarsius* and *Hapale.* Within the tympanum it occupies a bony canal which courses forwards and finally upwards. At the apex of the petrous bone, near the posterior clinoid process, the artery enters the cranial cavity, deep to the dura mater. It follows the usual course through the cavernous sinus lateral to the hypophyseal fossa, running forwards and then again turning upwards to pierce the dura. From its cavernous portion it supplies well-marked hypophyseal branches which ramify on the upper surface of the pituitary gland. Immediately after perforating the dura it gives off the ophthalmic artery, and thereafter it shortly ends by dividing into anterior and middle cerebral arteries.

The *ophthalmic artery,* arising as in *Tarsius* and *Homo* (Tandler's first category), accompanies the optic nerve into the orbit, lying infero-laterally thereto. In the orbit it turns medially, crossing over the optic nerve, and ends by dividing into frontal and nasal branches. Its main branches are lachrymal, retinal, a plexus of undulating branches surrounding the ciliary ganglion whence emerge the long and short ciliary vessels, and the supraorbital, which accompanies the frontal nerve to the supraorbital notch, finally ascending on the frontal region and anastomosing with the frontal and superficial temporal arteries. Ethmoidal branches are also present as well as palpebral arteries. An anastomotic circle is formed at the corneo-scleral junction. This is derived from the bifurcation of a fair-sized artery which courses over the upper surface of the eyeball just dorsal to the upper edge of the lateral rectus, to which it supplies twigs. It also supplies the levator palpebrae and the conjunctiva of the upper lid.

Circle of Willis (fig. 60): The bifurcation of the internal carotid into anterior and middle cerebral arteries occurs in the usual way, the former turning medially to enter the great longitudinal fissure between the hemispheres, the latter proceeding obliquely laterad in the Sylvian fossa and thence, after supplying a branch to the orbital aspect of the hemisphere, along the

lateral surface of the hemisphere, the major part of which is supplied by its half dozen or so branches. There is no anterior communicating artery, but the anterior cerebrals of the two sides unite soon after they enter the longitudinal fissure, after giving off a longitudinal vessel on the orbital surface of the hemisphere parallel to the medial border of the olfactory tract. The anterior cerebrals remain united until they gain the upper surface of the corpus callosum. Five millimeters behind the genu they again separate and are continued along the medial aspect of the corresponding hemisphere as far as the splenium. At this point they turn sharply dorsad, continuing thus halfway to the superior margin of the hemisphere. At this point each artery exhibits a kink, partly embedded in a deep dimple on the surface of the hemisphere. Emerging from this the artery takes a more oblique course upwards and backwards, breaking into terminal twigs which ramify caudally to within 5 mm. of the occipital pole. The more dorsal twigs transgress the superior border on to the upper margin of the superior surface of the hemisphere.

A posterior communicating artery is present on each side, linking the internal carotid with the vertebral system, forming thus a circulus arteriosus of Tandler's type I, i.e., resembling that of *Tarsius, Hapale,* and *Homo.* The circle is completed caudally by the diverging posterior cerebrals (profunda cerebri of Tandler), but there is some asymmetry in the example studied. This may be better appreciated by reference to fig. 60

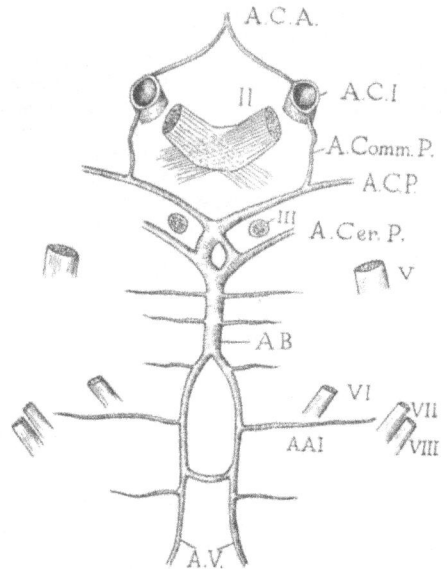

FIG. 60. Circle of Willis from below with related cranial nerve-roots.

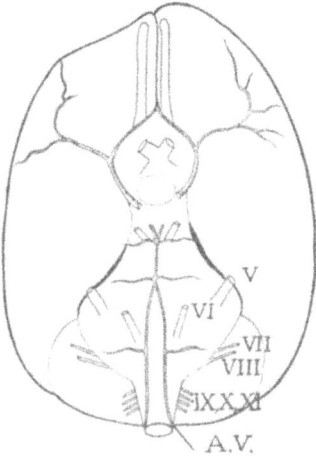

FIG. 61. Basal view of the brain showing distribution of cerebral arteries.

than by any lengthy description. The cross anastomosis between the two vertebrals should be noted halfway between their entry through the foramen magnum and the point of their union to form the basilar.

5. ARTERIES OF THE ANTERIOR EXTREMITY (fig. 62)

Subclavian Artery and its Branches

The subclavian takes the usual course, differing on the two sides, across the anterior aspect of the first rib in relation to the scaleni, the ventral member of which group divides the artery into the usual three parts. From the first part, i.e., medial to the scalenus insertion, arise the thyroid axis and vertebral arteries, the last-mentioned, however, almost under cover of the scalenus tendon. From the second part spring the superior intercostal and an ascending cervical artery.

The *thyroid axis* is a short, stout, forwardly-directed vessel which after a few millimeters breaks up into three divisions which are the suprascapular, internal mammary, and a short continuation of the main trunk which shortly bifurcates into transverse cervical and inferior thyroid branches.

The *suprascapular* crosses the scalenus ventralis and phrenic nerve transversely, entering the posterior triangle of the neck as in Man and *Nycticebus* (Davies)— not as in *Tarsius*. Passing under cover of the trapezius, it inclines caudally and crosses the cranial border of the scapula via the suprascapular notch accompanied by the nerve. It does not, however, follow the nerve to the infraspinous fossa, being lost in supplying muscular branches to the supraspinatus.

The *internal mammary* is a large vessel arising opposite the supascapular. Its arches over ventrally and turns caudad dorsal to the sterno-clavicular joint, then

courses along the lateral margin of the sternum within the thorax, finally bifurcating into anterior epigastric and musculo-phrenic at the joint between the sternum and the sixth costal cartilage. The terminal vessels have the usual relations and distribution. Laterally the internal mammary gives off intercostal branches to the anterior four intercostal spaces, besides twigs to the mediastinum and the comes nervi phrenici.

The *transverse cervical* courses dorsally parallel with the suprascapular but more anteriorly; it divides into superficial cervical and dorsal scapular branches, the former ascending beneath the edge of trapezius and the latter inclining towards the vertebral border of the scapula beneath levator scapulae and rhomboideus muscles.

The *inferior thyroid* is larger on the left than on the right. The left vessel is chiefly concerned in supplying several large branches to the thymus. The ascending portion and the whole vessel on the right are lost in supplying the cervical part of the oesophagus and trachea, leaving scarcely any twigs for the thyroid gland, which depends almost entirely upon the superior artery for its nourishment.

Vertebral Artery

This is a large vessel springing from the dorsal aspect of the subclavian and turning almost immediately craniad and somewhat medially to enter the vertebrarterial canal of the sixth cervical vertebra. It re-

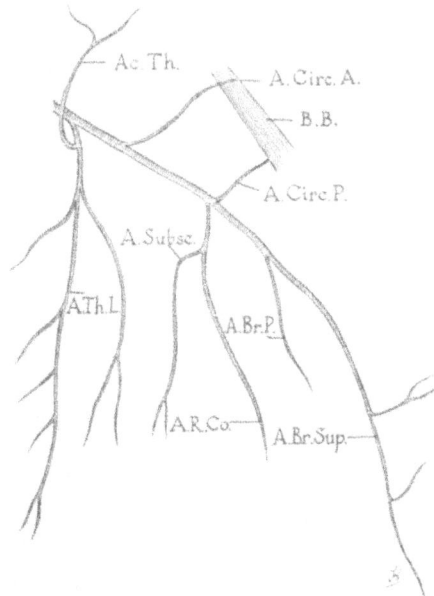

FIG. 62. Arteries of the left axillary and brachial regions.

ceives, however, a communication from the superior intercostal. Proceeding in the usual way through the vertebrarterial formina of the other cervical vertebrae, it turns laterally and anteriorly through the foramen in the atlas, winding thence dorsally and medially on the anterior edge of the dorsal arch of the bone, in the floor of the suboccipital triangle, in the normal relation to the first cervical nerve. Piercing the dorsal atlanto-occipital membrane and dura mater, it enters the skull through the foramen magnum, then, winding round the medulla, proceeds between the anterior rootlet of the first cervical nerve and the hypoglossal nerve fila. It gives off branches to the bulb and also the internal auditory arteries to the labyrinth, and then unites with its fellow at the medullo-pontine level, forming the basilar artery. This courses in the median line ventral to the pons, giving off small paired pontine twigs. At the anterior border of the pons it divides unequally as shown in fig. 60, the resulting four vessels being a pair of large superior cerebellar arteries and the rather smaller posterior cerebrals, the latter aiding in the completion of the circle of Willis. The cerebellar and cerebral branches are separated by the oculo-motor and trochlear nerves.

The *superior intercostal artery* proceeds across the neck of the first rib, then gives off the intercostal vessel to the first space and another branch which communicates with that in the second space. A communication is also made with the vertebral and a profunda cervicis is given off which ascends deep to the complexus to anastomose with segmental vessels from the vertebral and finally with the occipital.

The *ascending cervical* is a small vessel proceeding medially as far as the lateral border of the longus colli. On reaching this it turns sharply craniad, coursing along the groove between the longus colli and scalenus ventralis. This vessel is not, as in Man, a branch of the thyroid axis, but arises independently from the subclavian.

Axillary and Brachial Arteries (fig. 62)

The third part of the subclavian artery is continued first as the axillary artery and then, after passing the lateral edge of the teres major, the brachial, which remains single as far as the level of the deltoideus insertion, where it divides into brachialis superficialis and brachialis profunda, i.e., more distally than in *Hapale* (Bayer, 1892; Müller, 1904; Manners-Smith, 1910; Beattie, 1927) or *Aotes* (Bayer) but more in agreement with average conditions in the Cebidae.

Prior to this division the following are given off: (1) a short, stout stem which forms the common origin of a thoraco-acromialis and lateral thoracic arteries. It is given off from the concavity of the axillary, but the thoraco-acromialis turns cranially, ventral to its parent vessel, and divides into branches supplying the subclavius muscle and another turning laterally towards the acromion, supplying twigs to the deltoid. The

lateral thoracic is a large vessel with a long course along the lateral thoracic wall on the surface of serratus ventralis. It gives off at least three pectoral rami to the pectoralis major and minor, several alar thoracic branches which ramify in the axillary fat and lymphnodes and terminal twigs to the muscles of the thoracic wall.

(2) Two humeral circumflex arteries, each arising independently. The ventral one is a slender vessel arising a few millimeters past the origin of (1). It proceeds beneath the deltoid and across the short head of the biceps to ramify over the neck of the humerus. The much larger dorsal or posterior circumflex proceeds to the short head of biceps, proceeding thence through the quadrilateral space.

(3) A large vessel springing from the axillary at the same level as the posterior circumflex but on its concave side. It is a common stem which gives rise to the subscapularis and the collateralis radialis. The former takes the usual course along the lateral border of the subscapularis muscle alongside the long subscapular nerve. It reaches the posterior angle of the scapula and gives off the dorsalis scapulae branch, which crosses the cranial border of teres major to gain the dorsum scapulae, giving off branches to that muscle and to the infraspinatus.

The radial collateral artery accompanies the radial nerve to the back of the arm and then follows it into the forearm. It gives a large branch which supplies the long and lateral heads of triceps before proceeding to the distal part of the limb.

The *brachialis superficialis* continues superficially along the brachium to the medial side of the biceps superficial to and parallel with the median and ulnar nerves. Above the elbow it gives off muscular branches to coraco-brachialis, biceps, and brachialis anterior, and also the nutrient artery to the humerus and an arteria transversa cubita, which courses between brachialis anterior and the bone.

It enters the forearm superficial to the biceps tendon and becomes the radial artery. It divides into superficial and deep branches, the former being continued into the palm as the superficial volar artery. The deep branch could not be traced distally on account of the failure of the injection to penetrate its terminal parts.

The *deep brachial artery* courses in relation to the median nerve, following it to the elbow, where it accompanies the nerve to the forearm through the entepicondylar foramen of the humerus. In the depths of the antecubital fossa, it bifurcates into ulnar and interosseous arteries, and at the same time gives off a volar ulnar recurrent artery which ascends between pronator teres laterally and brachialis with coracobrachialis longus medially. It also gives rise to a dorsal ulnar recurrent which proceeds proximally between the medial condyle and the triceps tendon.

The ulnar artery thereafter inclines postaxially towards the ulnar nerve, which it then accompanies to the wrist, finally entering the palm, where presumably it assists the superficial volar in forming an arch whence the digital arteries take origin. It also supplies a dorsal branch which passes deep to the tendon of flexor carpi ulnaris to reach the extensor surface of the forearm.

The interosseous artery is as large as the ulnar and soon divides into dorsal and volar branches, which have the usual relations with the interosseous membrane and adjacent muscles. The dorsal interosseous gives off, as soon as it gains the extensor surface of the forearm, a large recurrent branch, which, after supplying the supinator muscle, follows its ulnar border towards the medial condyle, where it assists in the formation of the anastomosis round the elbow joint.

Collateralis radialis: As already mentioned this large vessel accompanies the radial nerve around the dorsum of the humerus, passing between the lateral and short heads of the triceps, and then proceeds towards the elbow between brachialis and brachio-radialis. It proceeds to the extensor surface of the forearm, running superficially, and presumably, as in *Tarsius,* ends on the dorsum of the hand in supplying the dorsum of the digits.

6. ARTERIES OF THE PELVIS, PELVIC GIRDLE, AND PELVIC LIMB (fig. 63)

The common iliac artery is 10.8 mm. long. It bifurcates into external and internal iliacs, the former having a course of 14.5 mm. to the point where it enters the thigh.

Internal Iliac Artery

This is 4.5 mm. long and emerges from the common stem at a very acute angle. Its course is continued as the hypogastric artery around the lateral wall of the pelvis, becoming obliterated after the first few millimeters, and continued as a fibrous cord not traceable, however, so far as the umbilicus. Obliteration commences on its cranial wall sooner than caudally, where a fine channel remains patent; along this blood is able to pass to the superior vesical artery, which springs from this wall of the vessel about halfway round the pelvic wall.

After giving off the hypogastric, the internal iliac continues posteriorly as a single trunk for some 1.5 mm. before bifurcating further into elongated vessels coursing almost parallel with each other. The lateral is the pudic and the medial the gluteal. There is no common pudendo-obturator trunk such as springs from the ventral part of the internal iliac in prosimians (Rau and Rao, 1930; Davies, 1947), since the obturator is, as in Hapalidae and all Cebidae except *Ateles,* a branch of the external iliac (Popowski, 1895). The pudic, after giving off a small inferior vesical vessel to the neck of the bladder and prostate and middle haemorrhoidal to the hindmost part of the rectum, proceeds in the usual manner, posterior to the pyriformis, through the great sciatic foramen to the buttock, returning to the

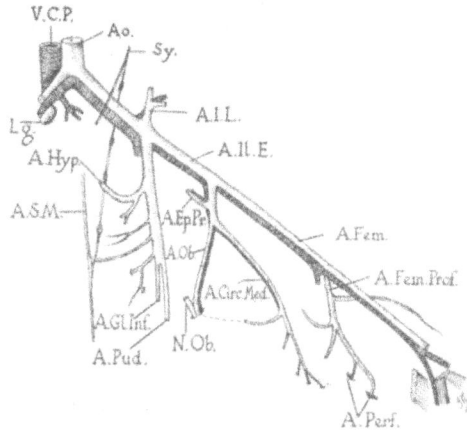

FIG. 63. Iliac vessels with their ramifications on the left side.

perineum *via* Alcock's canal and terminating, after supplying the usual perineal branches, as the dorsal artery of the penis. It supplies vessels to the anal canal and urethral bulb, following on the whole therefore the pattern described by Popowski (1895) for *Hapale penicillata.*

The gluteal artery sinks through the great sciatic foramen anterior to the pyriformis between the first and second sacral nerve-roots, meanwhile having given off a slender vessel which runs parallel with and medial to the pelvic part of the pudic artery. This seems to serve as a sciatic artery as it finally leaves the pelvis posterior to the pyriformis. There are no true lateral or middle sacral arteries. Segmental vessels entering the ventral sacral foramina anterior to the corresponding emergent nerve-roots are given off from the gluteal to the first two foramina and from the sciatic to the rest. The gluteal also provides a transverse branch which passes medially through the fibres of the pyriformis emerging on its medial side to unite with its fellow and then form a large median sacral artery, which proceeds posteriorly to become the caudal artery. Lateral caudal arteries seem to be derived from a backward continuation of the sciatic from the point prior to where it leaves the pelvis.

External Iliac Artery

As shown by Popowski and by Manners-Smith in *Hapale,* the external iliac gives off the ilio-lumbar artery at the same level as the internal iliac. After a short course it divides into iliac and lumbar branches. The latter runs forwards and joins in the longitudinal anastomosis in the loins (see p. 86). The iliac branch proceeds caudally along the edge of the ilium and anastomoses with the circumflex iliac. It is accompanied by the lateral femoral cutaneous nerve. At the point where the external iliac passes through the femoral arch it gives off the deep epigastric, circumflex iliac, and obturator branches. The deep epigastric has the usual relations to the neighboring fasciae and to the ductus deferens, finally proceeding forwards in the abdominal parietes to anastomose with the hinder intercostals and superior epigastric from the internal mammary artery. It supplies a cremasteric branch. The deep circumflex iliac has already been discussed.

The obturator commences as a large vessel, but suddenly narrows after giving off the medial femoral circumflex. Alternatively it may be regarded as a branch of the latter, which springs, as in *Hapale* and all Cebidae except *Ateles* (Popowski; Manners-Smith), from the external iliac instead of from the femoral. Crossing the posterior part of the psoas muscle, it inclines ventrally and passes into the thigh through the obturator membrane, in company with the obturator nerve which lies on its caudal side. It supplies iliac and pubic branches while still within the pelvis, and the lateral and medial terminal branches in the thigh.

The medial circumflex, as in *Hapale* and in all Cebidae except most examples of *Ateles* (Popowski; Manners-Smith), arises from the termination of the external iliac. It proceeds beneath the inguinal fascial arch into the thigh where it breaks up into an obturator branch (which takes a recurrent course to anastomose with the obturator artery) and a series of rami supplying the individual adductor muscles, terminating finally in the adductor magnus.

Femoral Artery

The external iliac is continued as the femoral artery after passing the groin. It enters Scarpa's triangle lateral to its companion vein and proceeds superficially at first, then beneath the sartorius into the adductor (Hunter's) canal, as in Man, but instead of passing entirely through the adductor magnus to become the popliteal, it divides into popliteal and saphenous arteries as in most subhuman Pithecoidea. Approximately 8.5 mm. from its commencement the femoral gives off from its deep aspect a large vessel which affords a common origin to the lateral circumflex and profunda. The former proceeds deeply and laterad shortly to divide into ascending and descending terminal branches, after giving an offset to the ilio-psoas. The latter courses just deep to the edge of vastus lateralis as far as the knee. The ascending branch proceeds proximally between the origin of the rectus and insertion of glutaeus medius, supplying both as well as the sartorius, but can scarcely be described as an a. circumflex ilii such as Bang (1936) mentions in *Lagothrix*.

The profunda femoris takes the usual course, deeply supplying the adductors and a medullary artery to the femur, terminating as a couple of aa. perforantes to the hamstring compartment. In *Hapale* only one perforating artery occurs, while in Cebidae several (up to four in *Ateles*) may be present, with the ex-

ception of *Saimiri*, where none was found by Manners-Smith. Bang does not specifically refer to aa. perforantes in *Lagothrix*, but mentions that the profunda supplies the hamstrings, hence their presence may be inferred.

Saphenous Artery

This is the larger of the two terminal divisions of the femoral. It takes the usual course, accompanied by the great saphenous vein across the medial aspect of the knee, passing first deep to the insertion of sartorius and then superficial to that of gracilis. Above the knee, it supplies the arteria suprema genu. The artery continues down the medial side of the leg, in close relation to the tibia. As in *Hapale*, it divides into an anterior and posterior branch at the junction of the upper and middle thirds of the leg, the latter being the posterior tibial. At the junction of the middle and lower thirds the anterior portion again divides into a larger anterior and smaller posterior division. The former turns forward deep to the tibialis anterior. As far as could be demonstrated this vessel was distributed as in *Lagothrix* as outlined by Bang. The smaller posterior division becomes the deep dorsalis pedis artery and presumably supplies perforating metatarsal branches as in *Hapale*.

Popliteal Artery

This enters the popliteal space through the lower part of the adductor magnus and disappears again between the two heads of the gastrocnemius and terminates by dividing into anterior tibial and peroneal arteries. Branches of the main vessel are articular to the knee and muscular to the heads of gastrocnemius and plantaris. There appear to be two genicular branches, medial and lateral, each of which divides into superior and inferior as is usual in *Hapale jacchus*, but there is a wide range of variation in this genus concerning the detailed arrangements of the four genicular vessels (*cf.* Beattie, who summarizes earlier findings). The muscular branches appear to correspond to the a. suralis described by Manners-Smith in *Hapale*

The peroneal artery continues down the leg, lying upon the deep flexor muscles. Distally it inclines towards the fibula and is somewhat covered by the peronaeus brevis muscle. After giving off a lateral calcaneal branch, it enters the sole, but its plantar distribution could not be verified in the specimen on account of the failure of the injection to penetrate beyond the ankle. At any rate the artery is more extensive than in *Hapale*, where Beattie sums up the results of his own observations and those of Popowski and Manners-Smith by stating that "it does not extend as a recognizable vessel beyond the middle of the leg." Both Cebidae and Hapalidae, however, agree in that the medial plantar artery is the larger and may be the sole supply to the foot, but the different genera appear to vary as to the source of the plantar arteries, i.e., as to whether they derive from the posterior tibial or peroneal.

IV. VENOUS SYSTEM

The cranial venous sinuses call for little remark beyond mention of the existence of a capacious inferior petrosal sinus connecting the cavernous sinus with the lateral sinus as the latter becomes continuous with the internal jugular vein. A large vena magna Galeni is present, continued backwards as the inferior sagittal sinus. The lower part of the lateral sinus is protected locally by a tonguelike spur of bone passing from the petrosal towards the exoccipital. It finally emerges through a separate compartment of the jugular foramen situated behind that transmitting the cranial nerves IX, X, and XI.

In the neck both internal and external jugular veins are present, the latter quite small. Caudally the external jugular, as in *Hapale,* divides near the clavicle sending a branch each side of the bone. The larger superficial branch (jugulo-cephalic trunk) (fig. 25) is joined by the cephalic vein from the arm. Deep to the clavicle this receives the other division of the external jugular, the combined trunk then immediately joining the subclavian vein. The internal jugular is then received, deep to the clavicle and medial to the first rib. Thus the innominate is formed. The two innominates join to form the precaval which, after a course of 4 mm., receives the azygos on its right side. After a further 2.5 mm. the precaval opens into the right atrium.

Beddard (1907) briefly dismisses the azygos system of the higher Primates by stating that there is but one azygos, the right, with, in cases, a left hemiazygos. He does not specify the species investigated, but claims to have examined a considerable number. *Callimico* falls into line with the majority, but Beddard's generalization is far too sweeping, since I have observed in *Tamarin midas* and in *Mico argentatus* a left azygos supplemented by a small anterior *right* hemiazygos.

The posterior vena cava is not unusual in its formation and disposition (*cf. Tarsius,* Hill, 1953), being formed dorsal to the aortic bifurcation by the union of two common iliac veins, each of which receives tributaries corresponding to the branches of the iliac arteries. It also receives near the median line a short branched tributary from the hollow of the sacrum. The femoral vein drains much of its blood from the internal saphenous which is formed on the medial side of the ankle coursing across the knee joint along the hinder border of the distal slip of the sartorius. Along the anterior slip of the sartorius insertion another large superficial vein proceeds, joining the preceding superficial to the belly of the sartorius. This drains the anterior and lateral parts of the leg distal to the knee joint.

The portal vein is formed by the confluence of the splenic vein with the anterior mesenteric, the posterior mesenteric already having emptied itself into the anterior mesenteric just prior to the junction of the latter with the splenic.

V. LYMPHATIC SYSTEM

Many of the lymphatic glands and nodes are brownish in color from accumulations of melanin pigment. This applies especially to those associated with drainage from the head and neck and body wall structures.

In the head region glands are located in relation to the parotid and submandibular salivary glands, a large one in front of the latter. In the neck there is a chain of glands along the carotid sheath with further agglomerations in the posterior triangle of the neck (cranially) and near the clavicle (supraclavicular group).

In the axilla is a large solitary apical gland and two distal chains of smaller nodes, separated by the intercosto-brachial nerve, which crosses the base of the axilla between the medial and lateral chains. No glands were seen in the antecubital fossa.

Inguinal glands are reduced to a single large oval pigmented gland occupying the femoral triangle.

There is a single large unpigmented flattened oval mesenteric gland situated on the anterior mesenteric artery at the site of origin of the right colic branch. The group of glands usually found both dorsally and ventrally in relation to the ileo-colic junction both in Hapalidae and Cebidae was not present in *Callimico.*

Brownish glands and nodes occur along the whole length of both thoracic and abdominal parts of the descending aorta. In the thorax they occur particularly along the dorsal wall of the hinder part of the vessel, forming a chain. In the abdomen there is a further concentration around the common iliacs and the neighborhood of the bifurcation.

On the whole the lymphatic arrangements are very primitive, especially as regards the visceral drainage. This is in strong contrast to arrangements in the Old World Primates where Beattie (1927a) has shown that the arrangement of the visceral lymphatic system shows little difference from the conditions in Man. It was not possible to confirm the well-known feature, first pointed out by Silvester (1912) as occurring in all platyrrhine monkeys examined, namely the connections between the visceral lymphatics and the venous system in the neighborhood of the confluence of the renal veins with the postcaval. It is hardly to be expected that *Callimico* would depart, in this respect, from normal platyrrhine standards.

NERVOUS SYSTEM

I. MENINGES (fig. 64)

The dura mater is a thick, firm, bluish membrane lining the cranial cavity and spinal canal. Only the cranial portion has been examined. Its relations show no unusual features; it is provided with falx cerebri and tentorium but no falx cerebelli. In the absence of the falx cerebelli *Callimico* agrees with *Hapale* and *Saimiri*. The falx cerebri is very narrow in front, where it virtually ends in a pointed apex. It broadens gradually behind, increasing to its attachment to the tentorium. The fixed border is considerably thicker than the free concave border and encloses the superior sagittal sinus. The dura is lightly attached over the orbital plates of the frontal bones and even in the temporal fossa is less firmly adherent than in higher Primates. The dura shows the usual relations to the hypophyseal fossa and neighboring parts.

Leptomeninges call for no special remark.

II. BRAIN (figs. 61, 65, 66)

In table 13 the measurements of the brain in *Callimico* are compared with those for *Tamarin, Hapale, Callicebus,* and *Saimiri,* i.e., with two members of the Hapalidae and two of the smaller representatives of the Cebidae.

TABLE 13

BRAIN MEASUREMENTS

(Capacity in cc.; weight in gms.; remainder in mm.)

	Callimico goeldii	*Tamarin midas*	*Hapale jacchus*	*Callicebus cupreus*	*Saimiri sciurea*
Cranial capacity	10.5 to 11	10	7.5	18 to 25	24
Brain weight	7.0	15	7.2	14.6	26
Height of cerebrum	30.4	40	30	39	50
Overlap beyond cerebellum	2.0	3.5	nil	2.0	7.5
Breadth of cerebrum	23.4	27	22.4	30.5	34
Height of cerebrum	18.5	20	17.9	24.5	26.7
Breadth of cerebellum	16.7	20	17.7	21.8	25.6

The brain of *Callimico* thus aligns itself, as regards proportions, alongside those of hapalids rather than the smaller cebids. The same remark applies equally to its shape and general conformation as can well be appreciated from the appearances of the bony braincase.

The ratio of brain weight to body weight is 1/39 compared with 1/29 in *Hapale* (Hrdlička, 1924) and 1/40 in *Hapale humeralifer* based on the data of Kennard and Willser (1941). Expressed as percentages, the following comparisons may be recorded: *Callimico,* 2.5 per cent; *Tamarin midas,* 6.3 per cent; *Hapale jacchus,* 3.7 per cent; *Oedipomidas oedipus,* 3.6 per cent (Hill, 1957). *Callimico,* therefore, has a smaller brain relative to its bodily size than any of the Hapalidae with which it has been compared.

The left cerebral hemisphere weighs 2.4 gms. and is slightly larger than its fellow, especially in length. The asymmetry also affects the outline as viewed from above, the right showing a smoother lateral curvature with the greatest lateral prominence somewhat more anteriorly than on the right hemisphere. The superior contour is convex but relatively flattened compared with cebid brains. On the other hand, it presents the narrow backwardly extended occipital pole of the brain of *Leontocebus* and *Tamarin.*

The supero-lateral surface presents the relatively lissencephalous character as in hapalid hemispheres, but, nevertheless, shows slight advances. The only sulci worthy of the name are the Sylvian and parallel sulci. The former is relatively longer than in the brain of *Hapale,* measuring 15 mm. from the Sylvian vallecula to its posterior extremity, which approaches within 5.5 mm. of the superior margin of the hemisphere. It is, as in *Hapale,* quite straight but is more oblique in direction.

The parallel sulcus, 6.5 mm. long, is a short, straight, deep fissure lying almost parallel to the Sylvian, but approaching it slightly at its upper end. It marks the temporal lobe somewhat nearer the antero-superior than the postero-inferior margin and falls short of the temporal pole by some 8 mm. Below the temporal and occipital lobes are separated by a well-marked concavity situated slightly nearer the occipital than the temporal pole.

Besides the two sulci described above there is a slight but distinct shallow impression above the Sylvian fissure in the position of the central sulcus of higher Primate brains. There is also a distinct dimple on the parietal region evidently representing the sulcus intraparietalis—this sulcus is stated by Connolly (1950) to be regularly present in *Oedipomidas,* though denied in *Oe. oedipus* by Elliot Smith (1902) who, however, mentions it in *Hapale penicillata.* In Elliot Smith's *Oedipomidas* there is a faintly marked depression in the position where the intraparietal might be expected to give off its posterior terminal offshoot.

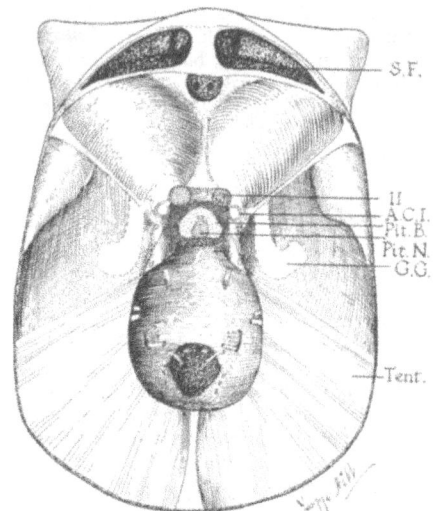

FIG. 64. Basis cranii interna. The tentorium cerebelli is *in situ.* The pituitary is also shown.

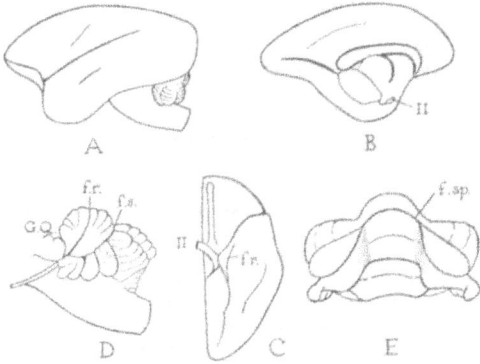

FIG. 65. Five views of the brain. A. Left lateral view of the entire brain. B. Medial aspect of the left cerebral hemisphere. C. Inferior aspect of the left cerebral hemisphere. D. Left lateral view of the cerebellum and medulla oblongata. E. Posterior view of the cerebellum.

On the orbital surface the small ellipsoidal olfactory lobe lies addressed to the hemisphere, leaving a distance of only 1.0 mm. between its cranial end and the frontal pole. The ribbon-like olfactory tract courses backwards some 1.25 mm. from the medial margin of the hemisphere, passing around the lateral edge of the tuberculum olfactorium to disappear in the vallecula Sylvii, finally gaining the pyriform lobe, which, as usual, is delimited by a distinct rhinal fissure. The olfactory peduncle leaves its mark on the hemisphere as a shallow groove, but there is no olfactory sulcus, neither could an orbital sulcus be determined lateral to the peduncle. Elliot Smith mentions an indistinct orbital sulcus in Oedipomidas, but none in Hapale penicillata, but Connolly mentions a deep depression in Hapale ? sp. and Oedipomidas.

On the mesial aspect the calcarine sulcus is the most conspicuous, being both long and deep. It extends backwards along the ridge dividing the mesial surface from the inferior surface of the hinder part of the hemisphere, from a point 2.5 mm. postero-inferior to the splenium of the corpus callosum to within 1.5 mm. of the occipital pole. It is relatively straight, unbranched, and occupied by the posterior cerebral artery. Deeply the calcarine extends, at any rate in its posterior part, more than half the thickness of the hemisphere. It is surrounded by a thick layer of grey matter which has a pinker color than the ordinary cortex, doubtless owing to a greater vascularity. Large vessels penetrate to the deepest parts of the fissure in the pial investment. No stria Genarii could be detected within this grey cortex. No paracalcarine is present.

A faint longitudinal depression is present in the expected position of the calloso-marginal sulcus. It extends from the mid-point of the length of the hemisphere backwards, almost gaining the calcarine fissure.

It does not appear to be due to vascular impression, a statement which is supported by the observation of Elliot Smith that a short intercalary or calloso-marginal is sometimes found in Hapale.

Inferiorly there is a well-marked sulcus collateralis. It takes the form of a broad, deep depression, somewhat more elongated on the right hemisphere than the left.

The hippocampal fissure, as stated by Elliot Smith in Oedipomidas, "presents those features which are common to all placental mammals."

The corpus callosum lies at the bottom of the great longitudinal fissure, 5 mm. from the surface. It measures 12 mm. long and 2.4 mm. thick at the splenium. It has the usual relations to the velum interpositum, septum lucidum, fornix, and lamina terminalis. Of the remaining commissures both anterior and hippocampal are small.

The optic thalamus has an antero-posterior diameter of 5 mm. but presents an appreciable postero-dorsal pulvinar. Geniculate bodies are comparable with those of Hapale (Woollard, 1926).

In the midbrain the corpora quadrigemina are well marked and of the usual shape. They are richly supplied with blood vessels, the anterior receiving three large branches from the posterior cerebral and the posterior a branch from the superior cerebellar.

The pons resembles that of Hapale.

The cerebellum (figs. 65, D, E.; 66) corresponds closely with that of Hapale as analyzed and depicted by Bradley (1904) who further maintains that, in this genus, the cerebellum differs materially from that of the lemurs, ranging itself convincingly with that of the higher Primates.

As in Hapale the cerebellum of Callimico agrees in the simplicity of the lobes and fewness of the folia. The general direction of the furrows is transverse, and there is no marked obliquity so frequently seen in other mammalian cerebella.

A few minor differences in shape between the cerebellum of Callimico and that of Hapalidae include a greater angularity of the lobus anterior both in lateral contour and in its transverse dorsal contour—especially

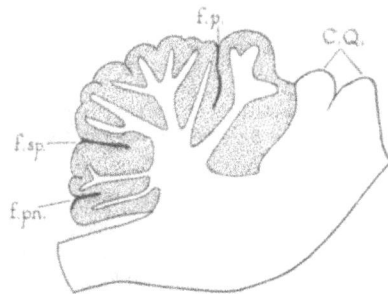

FIG. 66. Median sagittal section of the cerebellum.

as regards the former when compared with the organ in *Tamarin midas.* There is no clear division dorsally into vermis and lateral lobes by a paramedian fissure, but posteriorly there is a broad, shallow depression representing the paramedian sulcus, and this is deeper and more distinct than in *Hapale* or *Tamarin.* In these respects *Callimico* approaches near to *Aotes.*

Fissurae prima and secunda (suprapyramidalis) are well marked, especially the former, and take much the same course as in the cerebellum of *Hapale.* The lobus anterior is simple, with a triangular outline when viewed dorsally and convex in lateral view. It bears three folia dorsally and is separated from the lobus medius by the fissura prima, which crosses the summit of the cerebellum transversely. The lobus medius dorsally is vaguely dumbbell-shaped but ventrally it narrows towards the brachium conjunctivum. Dorsally it bears four minor folia bunched around a single large stem of the arbor vitae, while posteriorly two broader folia lie superior to the fissura secunda, which is deep, carrying vessels almost to the arbor vitae. An ansiform lobe is recognizable, though small; but the tonsil could not be recognized. In the lobus posterior the vermis bears a large pyramid and a somewhat thinner nodule, but the uvula is very compressed. Both pyramid and nodule are further subdivided into two minor folia by a transverse sulcus. The postnodular fissure is deep, about two-thirds the depth of the suprapyramidal, and, like it, contains a large vessel. Laterally there is a minute flocculus and a larger paraflocculus, borne on a slender pedicle attaching it to the nodule.

The cerebellar peduncles call for no remark, nor do the details of the medulla oblongata.

III. PERIPHERAL NERVOUS SYSTEM

1. CRANIAL NERVES

Olfactory Nerve

This consists of a thick bundle of loosely bound fila on each side, emerging from the apex of the olfactory bulb. The two bundles proceed forwards and slightly downwards in a single tube of dura enclosed in a stout-walled bony tube some 2.25 mm. in diameter and 5 mm. long. At the fundus of this bony tube the two nerves perforate the dura and pass through a single bony foramen each side to enter the nasal fossa, where they are distributed to the olfactory mucosa. Several small arteries accompany the olfactory fila within the dural tube. The arrangement appears to conform with that found in *Tarsius* and *Hapale.*

Optic Nerve

This has a length of 6.1 mm. of which 4.6 mm. lies in the orbit. The diameter just behind the globe is 2.1 mm., i.e., larger than in *Tarsius* despite the smaller eyeball. The nerve is surrounded by a firm dural sheath lined with arachnoid and pia. It emerges from

the globe on the nasal side and somewhat below the summit and proceeds posteriorly and medially towards the optic foramen. It is perforated below by the arteria centralis retinae.

Oculomotor Nerve (fig. 67)

This is large for the size of the animal and also in comparison with the size of the globe. Proceeding from the interpeduncular region of the midbrain, it enters the upper part of the cavernous sinus running forwards therein to the superior orbital fissure, where it enters the orbit within the fibrous annulus which gives rise to the ocular muscles. In the orbit it lies lateral to the optic nerve but closely adherent to it posteriorly. After giving off a branch which sweeps beneath the optic nerve towards the medial side to supply the medial rectus, it immediately afterwards gives another large (superior) branch which proceeds forwards within the cone of muscles. On the upper surface of the globe this breaks up into a leash of fine branches which enter the superior rectus and a single longer and thicker nerve which comes off medial to the preceding twigs and continues forwards to end in the levator palpebrae superioris muscle.

Meanwhile, the main nerve is continued forwards and shortly appears to enlarge. This is due to the adherence to its supero-medial surface of the large ciliary ganglion, which thus has no macroscopic radicles. Beyond the ganglion the nerve narrows abruptly and shortly divides into two more or less equal branches which proceed respectively to the inferior rectus and inferior oblique muscles, the latter nerve being the longer.

From the distal surface of the ciliary ganglion a large number of short ciliary nerves emerge. These proceed to the eyeball and enter its coats around the cribriform area.

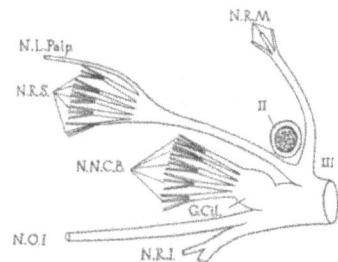

FIG. 67.   Left oculomotor nerve and ciliary ganglion.

Trochlear Nerve

This also a stout nerve. It arises from the dorsal surface of the midbrain immediately behind the posterior corpora quadrigemina. It sweeps around the brain stem, having the usual relation to the vessels, and enters the cavernous sinus. Its intraorbital course is normal and it ends in the superior oblique muscle in the usual way.

Trigeminal Nerve (fig. 68)

This emerges from the pons and has the usual large sensory and smaller motor roots. There is some interlacement of fibres belonging to the sensory division over the surface of the motor root as the latter travels across the deep surface of the trigeminal ganglion, thereby holding the motor root in position. The ganglion

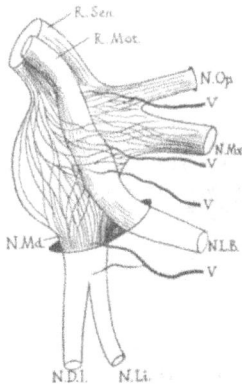

FIG. 68. Left gasserian ganglion from the deep aspect.

occupies the usual position on the petrous bone and has the normal relations to the dura. From its concave rostral border emerge the three divisions of the nerve, ophthalmic, maxillary, and mandibular, the two last mentioned subequal in size, the ophthalmic somewhat smaller. The motor root passes wholly to the mandibular division.

Ophthalmic and maxillary divisions proceed through the cavernous sinus in the usual way, the former entering the orbit *via* the superior orbital fissure and the latter *via* the foramen rotundum. Their further course and branches differ in no important particular from the arrangement in Man.

The maxillary division proceeds through the foramen rotundum to the spheno-maxillary fossa, where it divides into infraorbital and temporo-malar nerves, the latter almost as large as the former. They enter the orbit separately, the latter through a distinct foramen, and both run outside the periorbita. The infraorbital accompanies the corresponding artery in a groove on the orbital floor. This does not become roofed in by bone except at the orbital outlet, though a thin flange of bone partially protects it on the lateral side in the posterior half of the orbit. The nerve gives off a large bundle of fibres on the lateral side, but these do not emerge as a separate branch (anterior superior dental), rejoining the main stem for a short space first. After emerging through the infraorbital foramen the nerve

breaks up into large numbers of twigs supplying the lower eyelid, nose, cheek, and upper lip; a distinct twig emerges on the medial side and proceeds through a minute foramen over the root of the upper canine, presumably to enter the pulp canal of that tooth.

The temporo-malar nerve proceeds forwards and bifurcates before emerging on the cheek through two separate malar foramina, both of large size. The temporal branch was not identified. Malar branches supply the skin of the cheek lateral to the area supplied by the infraorbital nerve.

The mandibular division is the largest branch from the Gasserian ganglion. It proceeds through the foramen ovale lying deep to the lateral pterygoid muscle and behind the medial pterygoid and divides immediately into an anterior and a posterior trunk. The motor root is traceable solely to the anterior subdivision of which it forms the major part. It soon divides into branches supplying the temporalis (two large branches), masseter and lateral pterygoid muscles, leaving its small sensory component to continue as the buccal nerve which emerges in the interval between the two heads of the lateral pterygoid.

The posterior division gives off the auriculo-temporal nerve and then bifurcates into two stout branches, the lingual and inferior dental nerves. A few motor fibres must be included in the latter since the usual twig to the mylo-hyoid arises therefrom.

The otic ganglion lies deep to the main nerve, between it and the upper part of the belly of the tensor palati muscle. It is connected to the trigeminal *via* the nerve to the medial pterygoid prior to the entry of the latter into the deep aspect of the muscle.

The chorda tympani nerve joins the lingual branch of the mandibular nerve after escaping from the middle ear. It receives a filament from the otic ganglion.

Abducent Nerve

This nerve, large for the size of the animal, emerges from the brain-stem nearer the median plane and at a more posterior level than the fifth. Its origin lies lateral to the vertebral arteries just prior to their union to form the basilar and just craniad of the corresponding internal auditory artery. The nerve runs forwards beneath the forepart of the brain-stem to enter the cavernous sinus lateral to the carotid artery. It traverses the sinus in the usual way, entering the orbit through the superior orbital fissure between the heads of the lateral rectus muscle to whose medial surface it is applied, and in which it terminates.

Facial Nerve (fig. 60)

This emerges from the side of the brain stem in common with the auditory nerve at a slightly more cranial level than the preceding—in line with the internal auditory artery, which they accompany into the internal auditory meatus. The intraosseous course of the nerve could not be followed as it was not desirable to damage the bone; but this can scarcely differ to any appreciable degree from that in other Primates. The nerve emerges from the temporal bone *via* the stylo-mastoid foramen and breaks up at once into the usual leash of branches which enter the parotid gland, spreading out therein to emerge fanwise on the face. There is a temporal branch proceeding vertically in front of the external auditory meatus, and also well-defined cervical and occipital branches besides the forwardly directed stems.

Auditory Nerve (fig. 69)

From its origin, this consists of two distinct parts, the cochlear and vestibular nerves respectively. The cochlear nerve is the larger and enters the internal auditory meatus below the facial nerve and more anteriorly. It passes beneath the thick anterior extremity of the crista falciforme or transverse crest which divides the meatus into a superior and an inferior fovea. The vestibular nerve divides on entering the meatus into two branches, upper and lower, which sit astride the posterior, less well marked, part of the transverse crest. They separate at a wide angle, the upper division remaining closely associated with the facial nerve, but posterior to it. This (nervus utriculo-ampullaris) is the nerve to the utricle and also supplies the superior and lateral semicircular canals. The lower division (n. saccularis) is the nerve to the saccule and posterior semicircular canal. It is accompanied by the internal auditory artery.

Glossopharyngeal Nerve

Arising by rootlets from the hind-brain in association with the vagus, the two nerves emerge through the large, oval jugular foramen. Within the cranial cavity it is situated in front of the vagus, but, after leaving the jugular foramen in its own sheath of dura, it lies on its deep surface. The early extra-cranial course is deeply situated between the bulla and the middle part of the condyle, and is embedded in dense connective tissue which renders dissection difficult. It gives fine communications to the vagus. Only one ganglionic swelling,

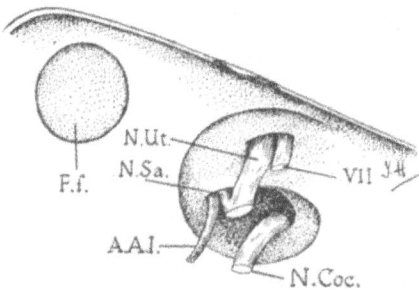

FIG. 69. Nerves, etc., related to the fundus of the left internal auditory meatus.

recognizable by its dense white appearance, could be detected in this part of the nerve.

In the neck it inclines forwards, taking the usual course around the dorsal border of the stylo-pharyngeus muscle, which it supplies, to enter the pharyngeal wall, supplying tonsillar branches and thereafter entering the tongue beneath the hyoglossus. Its tympanic branch pierces the posterior part of the medial wall of the bulla. Pharyngeal branches were noted.

Vagus (fig. 58)

Much larger than the preceding, this emerges through the jugular foramen immediately after receiving a contribution from the eleventh nerve, whose dural sheath it shares. Its trunk presents a medio-laterally compressed enlargement (ganglion trunci vagi). From this part of the nerve arise the pharyngeal and superior laryngeal nerves. The former joins the pharyngeal plexus and the latter divides into the usual external and internal laryngeal branches. The main trunk continues its course to the thorax in company with the carotid artery and internal jugular vein. Both the main trunk and the recurrent laryngeal branch present the usual asymmetries in respect of their relations to the great vessels. The formation of the plexus gulae and the final emergence of the right and left vagi in their new positions relative to the oesophagus and their final passage through the diaphragm call for no remark.

Accessory Nerve

This is formed by the union of bulbar and spinal portions, the latter taking origin from the first 5–6 cervical segments of the spinal cord by separate fila which unite to form a sagittally coursing trunk, which proceeds through the foramen magnum to join the bulbar fila springing from the sides of the medulla caudal to the rootlets of the vagus. The cervical fibres are destined to supply the sterno-mastoid and trapezius muscles, the bulbar fibres being distributed to alimentary musculature *via* the vagus, which they join at the base of the skull outside the jugular foramen.

Hypoglossal Nerve

A number of rootlets emerge between the pyramid and the olive. It was not possible in the state of the available material to confirm Fieandt's (1914) findings on *Cebus* that the rootlets do not emerge in a straight line, but this is certainly the case in *Hapale*, so that doubtless *Callimico* agrees. Other data in respect of this nerve recorded by Fieandt are the independence of the hypoglossal fila from those of the vago-accessorius complex and the first cervical nerve, to none of which does it give communications. The hypoglossal fila in *Cebus* are arranged into three groups of differing calibre. The oral group composed of very thin fila, a middle group of much thicker components and a caudal series of very strong development. Each group pierces the dura separately in *Cebus;* in *Hapale* this distinction is not obvious, but in *Callimico* the remnants of dura suggest multiple perforation. In *Cebus* the oral fila do not unite with the middle series until they are beyond the dura. A tributary of the basilar vein separates the middle and caudal fila in *Cebus*, *Hapale*, and *Callimico*.

Emerging through the condylar foramen the hypoglossal trunk takes the usual course in the neck, being at first very deeply placed between the condyle and the medial aspect of the vagus and deep cervical vessels. It winds round the ganglion trunci vagi, to which it is connected by fibrous tissue, then proceeds downwards and forwards between the carotid artery and the internal jugular vein. It winds round the posterior border of the digastric and the occipital artery, then above the hyoid bone to enter the tongue superficial to the hypoglossus. Before this it gives off a fine descendens hypoglossi which supplies the infrahyoid muscles, branches being traceable as far posteriorly as the thoracic part of the sterno-thyroid. An ansa is formed with the descendens cervicis from the cervical plexus. The ansa is located relatively caudad.

2. SPINAL NERVES

There are eight cervical, twelve thoracic, seven lumbar, and four sacral nerves. The usual arrangements prevail as regards bifurcation into dorsal and ventral primary divisions. The ventral primary divisions of cervical nerves 1 to 4 enter into the formation of the cervical plexus. The brachial plexus is contributed to by cervical nerves 4 to 8, and the first thoracic nerve. The remaining thoracic ventral primary divisions, except the last pair, constitute typical intercostal nerves. The last thoracic (subcostal) nerve, together with the ventral primary divisions of all lumbar nerves, enters into the lumbar plexus. The sacral (sacro-pudendal) plexus is formed by the ventral divisions of nerves S.1 to S.4 and the first coccygeal.

Suboccipital Nerve and Cervical Plexus (fig. 70)

The cervical plexus is essentially a series of loops between the ventral primary divisions of the anterior four cervical nerves owing to the splitting of each nerve (except the first), as it emerges from the intervertebral foramen, into ascending and descending parts, which anastomose with their neighbors to form the loops.

As regards the first cervical nerve, it should be mentioned that the ventral primary division is exceptional in its smaller size than the dorsal, which constitutes the *suboccipital nerve*. The latter emerges over the dorsal arch of the atlas, between it and the vertebral artery, and ramifies, as usual, in the suboccipital triangle. It appears in the fat medial to the obliquus capitis superior and the rectus capitis dorsalis major, or perforating the lateral fibres thereof. It gives motor nerves to all the suboccipital muscles including a twig

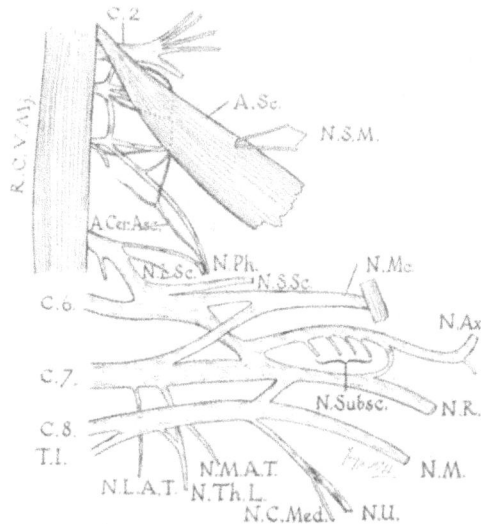

FIG. 70. Cervico-brachial plexus of the left side from the ventral aspect.

to the longissimus capitis (complexus). No cutaneous twig was found.

Branches from the cervical plexus consist of a deep and a superficial set. According to Cordier *et al.* (1936) the cervical plexus in *Hapale*—the only platyrrhine examined—is simpler in construction than in any other subhuman Primate studied, where the arrangement invariably resembles the most usual of the three types recognized in adult humans, i.e., an ansiform plexus with the branches arising from the trunks rather than the loops. In their figure of *Hapale* a loop is shown between C.1 and C.2, and an oblique connection between C.2 and C.3, apart from which branches of distribution are given off directly from the trunks. Bolk (1902a) in *Leontocebus* (= *Leontideus*) shows no loops whatever so that virtually there is no plexus-formation beyond the linkage of C.2 and C.3 and the contribution of C.3 and C.4 to the phrenic.

In contrast with these facts, *Callimico* shows definite loops between C.1, C.2, and C.3, but the distributional branches are derived solely from the trunks.

C.1 emerges between the rectus capitis lateralis and rectus capitis ventralis minor. Others emerge between the corresponding intertransversarii and then insinuate themselves between levator scapulae and scalenes dorsally and the rectus capitis ventralis major ventrally. All these muscles receive branches from one or more of the cervical nerves. A curious feature is the perforation of the omo-cervicalis, near its lateral border, by the nerve to the sterno-mastoid derived from C.3 and C.4. Bolk shows a close relation between the nerve and this muscle in *Leontocebus*, but not an actual perforation. The nerve to the trapezius is a long trunk carried onwards on the deep surface of the muscle after the parent nerve has supplied and perforated the omo-cervicalis and given branches to the sterno-mastoid. Thereafter it perforates the levator scapulae and courses obliquely mediad, then sagittally along the deep aspect of the trapezius near its origin. Its fibres are derived from segments C.3 and C.4.

The great auricular nerve is largely derived from C.2, which is a bigger nerve than C.3, which also contributes some fibres. The lesser occipital nerve is also from C.2 and C.3. The transverse cervical nerve and the descending supra-clavicular cutaneous branches were recognized. These are derived from C.3 and C.4. The phrenic has the usual composition, C.4 and C.5 in about equal degree, and takes the usual course ventral to the scalenus ventralis and within the thorax, crossing the subclavian artery on the right and the aortic arch on the left, thence ventral to the lung root to the cranial aspect of the diaphragm.

Brachial Plexus (figs. 70, 71)

The general arrangement of the nerve roots, trunks, and cords agrees with that found in *Hapale* (Sterzi, 1903), *Leontocebus* (Bolk) and *Tamarin* (personal observation), i.e., nerves C.5 and C.6, the latter being the larger, unite to form the anterior trunk; C.7 proceeds alone to form the middle trunk; C.8 and T.1 unite early to form a flat band over the neck of the first rib, constituting the posterior trunk. Each of these trunks bifurcates into a dorsal and a ventral division. The three dorsal divisions then unite to form a dorsal cord; the ventral divisions of the anterior and middle trunks unite to form the lateral or antero-ventral cord, while the ventral portion of the posterior trunk continues as a medial or postero-ventral cord.

From the nerve roots twigs are given to the prevertebral muscles and scalenes and from C.5 at least a fair-sized branch to levator scapulae. The suprascapular nerve is of large proportions and springs from the plexus more distally than in the higher Primates. It derives fibres from C.5 and C.6 and proceeds to the suprascapular notch and supplies the supraspinatus and infraspinatus. The nerve to the subclavius is also relatively large and wholly from C.5. It proceeds over the third part of the subclavian artery to enter the deep aspect of the muscle. The long thoracic nerve (of Bell) springs from the dorsal side of the root region of the plexus by the union of twigs from C.7 and C.8. (In *Leontocebus* Bolk indicates it from C.6, C.7, and C.8 as well as a small contribution from T.1.)

The chief branch from the lateral cord is the lateral anterior thoracic (pectoral) supplying the pectoralis major. It is a stout nerve derived chiefly from C.7 with contributions from C.8 and T.1. The medial cord gives off the smaller medial pectoral to the pectoralis minor. This derives fibres from C.8 and T.1 only. The posterior cord gives off at least four subscapular nerves,

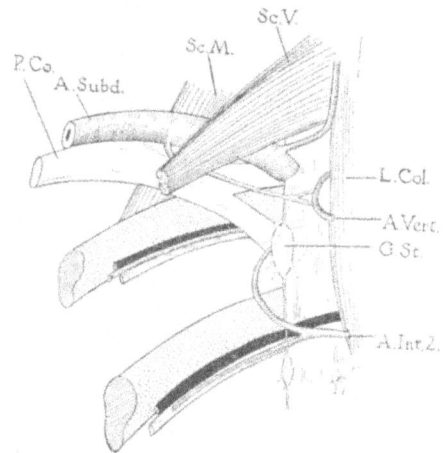

FIG. 71. Nerves and vessels at the cervico-thoracic junction.

besides the nerve to the latissimus dorsi. The upper series all end in the subscapularis. Thereafter the cord divides into radial and axillary (or circumflex) nerves, the latter giving off the nerve to the teres major before passing to the dorsal scapular region.

The lateral cord is continued as the *musculo-cutaneous nerve* and the main part of the median, which has a long lateral head and a short medial one from the medial cord, the discrepancy being similar to that observed in *Tamarin midas.* The musculo-cutaneous has the usual arrangement in the arm, supplying coraco-brachialis, biceps, and brachialis and terminating as a cutaneous nerve in the forearm. The branch to the brachialis is a long nerve coursing deep to the biceps parallel with the lateral border of the coraco-brachialis muscle. The *median nerve* shows the normal relationships to the brachial artery, coursing superficially parallel with the medial border of the biceps. Giving no branches in the brachial region, it accompanies the artery through the entepicondylar foramen and enters the antecubital fossa between the biceps and pronator-flexor origins. It gives off at once a leash of motor branches to the muscles arising from the medial epicondyle and further ones along its course in the forearm. Much reduced it continues over the wrist embedded in the flexor retinaculum. In the hand it delivers a bunch of branches to the muscles of the thenar eminence and then terminates in digital nerves to the palmar skin of the radial three and a half digits. A communication is made with the palmar cutaneous branch to the ulnar nerve.

As regards the thenar branch of the median nerve, Winckler (1930) drew attention to its acutely recurrent course in *Hapale,* the angle depending apparently on the relative divergence of the pollex. *Callimico* does not differ from *Hapale* in this respect.

The *ulnar nerve* is the largest derivative of the medial cord, which also gives off the medial cutaneous nerves of the brachium and antebrachium. Like the median, the ulnar gives no branches in the brachium. It enters the forearm behind the medial epicondyle between the two heads of the flexor carpi ulnaris and then proceeds along the medial side of the ulnar artery. Previous to this it is accompanied by ulnar collateral branches of the brachial. In the forearm a large dorsal cutaneous branch is given off, which becomes superficial after winding dorsally beneath the flexor carpi ulnaris. A palmar cutaneous branch crosses the wrist after piercing the deep fascia. The dorsal branch supplies the ulnar one-and-a-half digits and links up with the radial nerve. In the palm the main trunk divides into superficial and deep terminal branches. The latter supplies the hypothenar muscles, interossei and contrahentes as well as the medial lumbrical muscles. Its course is deep to the contrahentes.

The *radial nerve* (fig. 72) is a large trunk continued from the dorsal cord of the plexus. Taking the usual spiral course in relation to the shaft of the humerus and with normal relations to the triceps, it eventually appears on the lateral side of the elbow between the brachialis and brachio-radialis muscles. In the forearm it accompanies the radial artery and courses to the wrist under cover of the medial edge of the brachio-radialis and extensor carpi ulnaris. It does not lie upon the supinator, but passes more dorsally.

Immediately after passing the distal border of teres major the radial gives off a substantial short branch to the lateral head of the triceps and another longer one which proceeds distally along the deep face of the muscle, also supplying it (the long head is supplied while the nerve is still in the axilla). About the level of the middle of the humeral shaft two branches are delivered to the medial head of the same muscle, a short one on the lateral side and a much longer but very slender one which proceeds to the elbow and ends in the anconeus. In the distal third of the arm two lateral cutaneous branches are given off. These become superficial and proceed to the lateral side of the forearm. Passing medial to the brachio-radialis and extensor carpi radialis longus, the parent trunk now supplies individual branches to these two muscles and to the other muscles arising from the lateral epicondyle of the humerus. Other twigs are given to the supinator. The main nerve does not penetrate the supinator, but continues across it, or across the bone near the dorsal margin of the muscle. Shortly after, the stem divides into a larger cutaneous and a slenderer posterior interosseous branch. The former becomes superficial in the distal third of the forearm, by passing between the brachio-radialis and extensor carpi radialis longus tendons. It soon after divides into a dorsal cutaneous branch to the dorsum of the wrist and hand, and a terminal digital portion which proceeds on the dorsum of the hand supplying the radial border of the thumb and contiguous borders of the first, second and third interdigital clefts, as Kosinski (1927) found in *Hapale*, *Leontocebus*, and *Cebus*, the remainder being supplied by the dorsal branch of the ulnar nerve. According to Kosinski the wider territory supplied by the radial nerve is a more primitive condition and contrasts with the reduced distribution observed in the Pongidae and *Homo*.

The posterior interosseous nerve is very slender, but traceable as far as the wrist. It is purely motor and articular, supplying the deep extensor muscles and the wrist joint.

Thoracic Nerves (fig. 71)

The first thoracic nerve is involved, as already stated, in the brachial plexus. Its root is related closely to the stellate ganglion of the sympathetic chain. The second, apart from supplying a lateral branch which extends across the axilla as an intercosto-brachial nerve, which supplies cutaneous fibres to the medial aspect of the upper arm, behaves as an ordinary intercostal nerve. No connections were traced between the intercosto-brachial, or the parent nerve, with the brachial plexus.

The remaining thoracic nerves are distributed in the normal way as intercostal nerves.

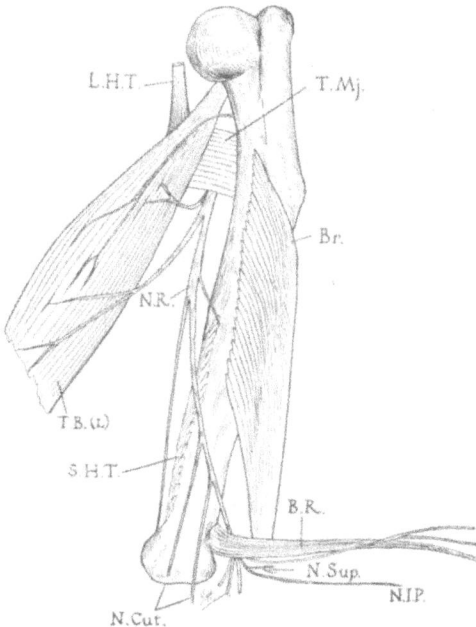

FIG. 72.   Course of the radial nerve in the right
brachium and antebrachium.

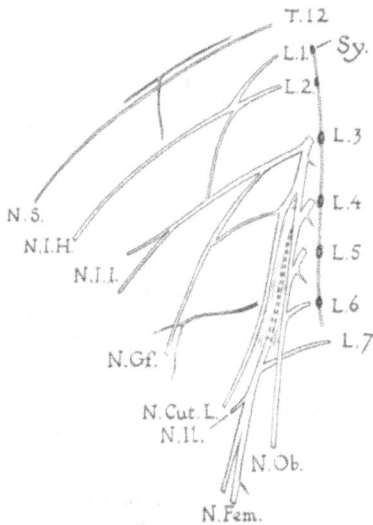

FIG. 73. Right lumbar plexus from the ventral aspect.

Lumbo-sacral Plexus (figs. 73, 74)

All seven lumbar nerves enter into the formation of the lumbar plexus. No communication was found between the subcostal nerve and L.1. The plexus is situated, for the most part, in the substance of the psoas magnus or between it and psoas parvus. The ventral primary divisions of L.1 and L.2 unite and send a communication to L.3. The conjoined stem formed by L.1 and L.2 is continued as the ilio-hypogastric nerve. The cranial part of L.3, after receiving the communication from the preceding nerves, is continued as the ilio-inguinal nerve. A thin nerve is also given off which is joined by another from the communicating band between L.3 and L.4. This union forms the genito-femoral nerve. Also deriving fibres from L.3 and possibly L.4 is the relatively large lateral femoral cutaneous nerve.

The femoral nerve ("anterior crural") is a large trunk formed within the substance of the psoas magnus by roots from L.3 to L.7. It supplies a branch to the iliacus, then continues into the thigh lateral to the femoral vessels.

The *obturator nerve* is also a large trunk. It lies lateral to the psoas magnus, parallel to its lateral border and covered at first by psoas parvus. Its fibres are derived from L.4 to L.7.

All the lumbar roots give twigs to the psoas muscles and to quadratus lumborum. In their course in the flank the trunks derived from the upper three lumbar nerves lie between the internal oblique and transversus abdominis muscles. They are accompanied by arteries derived from the segmental vessels.

The greater part of the last lumbar nerve proceeds posteriorly over the sacro-iliac joint, forming the lumbo-sacral trunk. The sacral and pudendal plexuses are quite distinct as Cordier *et al.* (1936) found in most lower Primates, and as figured by them in *Hapale*, i.e. there is no contribution from S.1 to the sacral plexus, which is comprised solely by the ramification of the lumbo-sacral cord, deriving its fibres solely from the last two lumbar nerves.

Ranke (1897) remarked that the formation of the sciatic plexus from three roots only was exceptional in the material examined by him, four being the modal number, the exception being *Cebus*, where the three roots all supplied fibres to both peroneal and tibial portions of the great sciatic. No hapalid was included in Ranke's series, but an isolated *Callithrix* (presumably a *Callicebus* or *Saimiri*) showed an extremely slender fourth root, which was interpreted as an intermediate (transitional) stage in the process of segmental shifting of the extremities. Reduction to two roots seems, therefore, to be an evolutionary change in the reverse direction.

The lumbo-sacral cord is primarily continued into the buttock, caudal to the pyriformis, as the great sciatic nerve, but gives off as collaterals the anterior gluteal nerve which supplies the deeper glutaeus muscles, and twigs to the pyriformis. Others are given off after entering the buttock, as described below.

Pudendal (Pudendo-coccygeal) Plexus (fig. 74)

This is composed of linkages between sacral nerves 1–3 and the coccygeal (or first caudal) nerve. The principal outcome of this is the pudendal (internal pudic) nerve which is derived mainly from S.1, but which receives a fair-sized contribution from S.2. This is not as shown by Cordier *et al.* in *Hapale*, where a contribution from the last lumbar is received peripherally and none from S.2. Another large derivative of this plexus is the lateral caudal nerve, chiefly from S.2, but with contributions from the remaining roots of the plexus. Collateral branches are given from S.1 to the

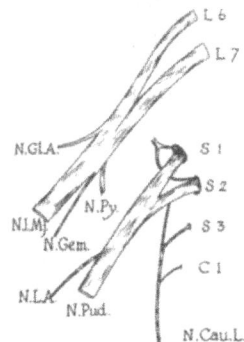

FIG. 74. Right sacro-pudendal plexus from the ventral aspect.

pyriformis and from S.1 and S.2 to the nerve to the levator ani, also inferior rectal nerves.

The pudendal nerve takes the usual course out of the pelvis behind the pyriformis into the buttock, returning thence to the perineum, where it courses along the inner surface of the obturator internus. It supplies twigs, across the ischio-rectal fossa, successively to ischio-coccygeus, pubo-coccygeus, and pubo-rectalis, with sphincter ani. It is then continued ventrally, supplying ischio-cavernosus and bulbo-cavernosus, to become finally the dorsal nerve of the penis.

Distribution of Nerves in the Pelvic Limb (figs. 75, 76)

Femoral Nerve

Entering the thigh lateral to the femoral artery, this nerve gives first a slender twig to the ventral portion of the pectineus, and also twigs to ilio-psoas. Then on its lateral side springs a short trunk which divides almost immediately into a branch to the sartorius and a long cutaneous branch which follows the great saphenous vein beyond the knee. The remainder of the main nerve courses medially and deeply between the rectus and vastus lateralis on the one side and vastus medialis on the other. It is concerned solely in the innervation of the parts of the quadriceps, giving branches successively to the rectus, vastus lateralis, vastus intermedius, and vastus medialis: All three last mentioned are traceable to the knee. No medial or intermediate cutaneous branches were revealed, but the large saphenous nerve doubtless serves their territory. A subsartorial plexus was found with contributions from femoral, saphenous and obturator nerves. Fine branches were also traceable to the wall of the femoral artery.

Obturator Nerve

This is a thick nerve appearing in the thigh deeply between the adductors, the pectineus and adductor longus lying laterally and adductor brevis and adductor magnus medially. The trunk breaks up almost at once into branches to the individual adductor muscles, but also contributes to the subsartorial plexus. A large branch is first given to the gracilis which enters the muscle near its ventral border. Smaller branches are given off to the pectineus and adductor longus, larger and longer ones to adductor brevis and adductor magnus. Deeply also a branch is supplied to obturator externus. The nerve is related to the medial circumflex artery, which passes dorsal to it.

Distribution of Derivatives of Sacro-pudendal Plexus in the Pelvic Limb

The pudendal nerve, during its course through the gluteal region, gives muscular branches to the femorococcygeus, as shown by Appleton (1928) in *Hapale*. The muscle termed by Appleton caudo-femoralis, though

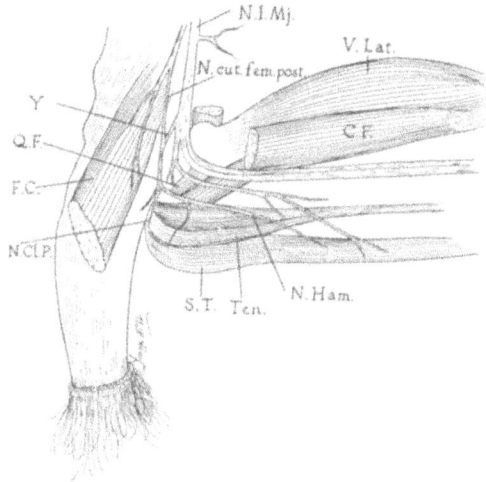

FIG. 75. Dissection of right gluteal and adjacent part of femoral region to show distribution of nerves.

present in *Hapale,* is wanting in *Cebus,* as is also the flexor cruris caput dorsalis. *Callimico* seems to agree rather with *Cebus.*

Between the pudendal and great sciatic nerves, the posterior femoral cutaneous (small sciatic) nerve decends to the thigh. This is intimately linked at first to the pudendal, but more distally receives a communication from the great sciatic, evidently the homologue of the nerve 'Y' found by Appleton in *Cebus.* It joins the main stem in a Y-shaped junction. The nerve then follows the groove between the superficial gluteal muscle and the hamstring musculature. As usual, tenuissimus lies dorsal to this nerve. As in Man, the posterior cutaneous gives recurrent nervi clunium posteriores and continues down the back of the thigh as a cutaneous nerve.

As in most mammals, there are separate nerves to obturator internus and quadratus femoris (together with the associated gemelli).

After the nerve "Y" has been given off, the nerve to the hamstrings separates itself from the dorsal border of the great sciatic nerve. This proceeds dorsad and supplies branches to semitendinosus, biceps, and semimembranosus as well as the ischio-condyloideus portion of adductor magnus.

The main nerve turns over the dorsal border of the great trochanter at its root and continues undivided deeply down the back of the thigh. A distinct groove on its surface can be readily seen dividing it into its tibial and peroneal portions, the former located more dorsally. The final separation of these two occurs at the junction of the middle and lower thirds of the femoral shaft, or slightly more distal.

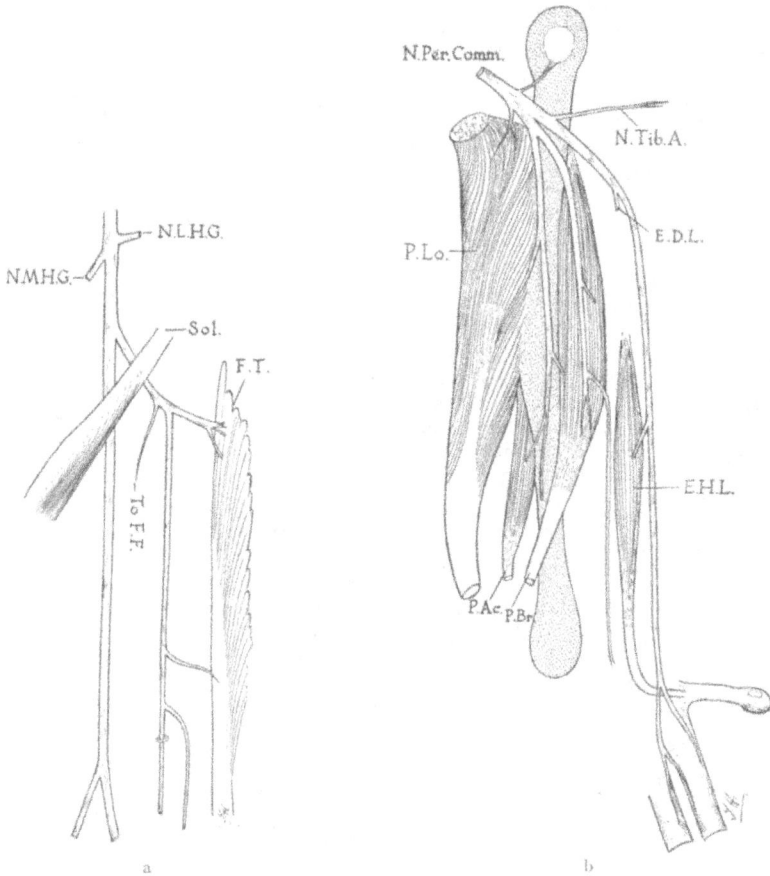

FIG. 76. Two figures to illustrate the distribution of nerves in the crural region and dorsum of the foot.

The peroneal division keeps to the lateral side of the popliteal fossa and proceeds deep to the biceps tendon into the substance of the uppermost part of the origin of the peronaeus longus. Here it lies against the shaft of the fibula somewhat distal to the neck of the bone. It divides here into four branches, a small one to the origin of peronaeus longus, and three larger ones, the most anterior being the anterior tibial nerve, the other two constituting a musculo-cutaneous. The anterior tibial nerve gives off at once a large branch to the uppermost part of tibialis anterior, then turns distally between the extensor digitorum and tibialis anterior and afterwards between the latter muscle and the extensor hallucis longus. It gives further muscular branches to all three muscles and then crosses the ankle joint to end as a cutaneous nerve in the foot, where it supplies a dorsal branch to the skin of the

foot and ends in digital nerves to the second inter-digital cleft.

The two parts of the musculo-cutaneous are long slender nerves one of which descends between peronaeus longus and peronaeus brevis, while the other proceeds between the latter and the anterior peroneal septum. It eventually pierces the septum and emerges in the superficial fascia between the peroneal and extensor groups of muscles to become a cutaneous nerve. Muscular twigs are given from both divisions to all the peronaei. The arrangement exactly agrees with Winckler's (1934) findings in *Hapale* and *Cebus*, which differ therein from *Ateles*.

The tibial (medial popliteal) nerve proceeds down the middle of the popliteal fossa between the two heads of the gastrocnemius. It supplies muscular branches to both bellies of this muscle.

The main nerve continues down the calf between the superficial muscles (gastrocnemii, plantaris, soleus) and the deep flexors of the foot and toes. It divides beneath the flexor retinaculum at the ankle into medial and lateral plantar nerves. Just prior to passing deep to the tendon of origin of the soleus, it gives off a large branch which becomes the nerve accompanying the peroneal artery. This gives off shortly the nerves to flexor tibialis and flexor fibularis, then proceeds as a slender filament along the surface of the tibialis posterior. When this muscle gives place to its tendon, a branch accompanies the tendon to the sole, but a finer filament continues the course of the parent nerve on to the interosseous membrane, which it soon perforates to end on the extensor side of the ankle joint.

A single large sural nerve is given off from the tibial nerve. This receives no ramus communicans fibularis and, after running deeply between the two bellies of the gastrocnemius on to the tendo Achillis, proceeds obliquely, accompanying the short saphenous vein to become cutaneous to the fibular side of the dorsum of the foot and lateral aspect of the fifth toe and contiguous borders of the fourth interdigital cleft. It forms communications with more distal branches of the n. tibialis, including the lateral plantar nerve. This is essentially as described by Ssokolow (1933) in *Hapale jacchus* and *Cebus capucinus*.

Of the two plantar nerves, the lateral is the larger. Both are long nerves, owing to the elongation of the tarsal and metatarsal region. Both nerves give off immediately, i.e., at the site of their own divergence from the posterior tibial (the separation into four distinct nerves before the divergence is very distinct within the connective tissue sheath of the parent nerve), fair-sized cutaneous branches supplying the heel region and sole. Others are given to neighboring muscles, to the thenar mass by the medial nerve and to the hypothenar mass by the lateral. Digital branches from the medial nerve continue the course of the main nerve to the distal part of the metatarsal region whence they diverge to supply the medial side of the hallux and adjacent sides of hallux and index and index and medius.

The lateral plantar completes the digital supply, but the digital branches arise from the deep, not the superficial division of the nerve. The main nerve proceeds distally accompanying the artery between flexor digitorum brevis and flexor accessorius. At the distal border of the latter muscle, it bifurcates into superficial and deep divisions. The superficial supplies the lumbricals, or at least two lateral ones. The deep division supplies contrahentes and all the interossei, as well as flexor brevis digiti minimi.

3. AUTONOMIC NERVOUS SYSTEM (figs. 63, 71)

The sympathetic chain commences in the neck with the superior cervical ganglion, a firm, grey, oval body 2.5 mm. long and 1.0 mm. thick. It has the usual connections with the cranial nerves and the carotid plexus. From its caudal end a fine nerve proceeds which exhibits no further ganglionic enlargements in the neck. A large ganglion stellatum, presumably representing the fused inferior cervical and first thoracic ganglia, lies opposite the dorsal end of the first intercostal space, its cranial end overlying the root of the first thoracic spinal nerve as it passes obliquely across the neck of the first rib to meet the last cervical in the formation of the brachial plexus. The stellate ganglion measures 3.0 mm. long and 1.3 mm. thick; it receives rami communicantes and gives off numerous branches of distribution.

The remaining thoracic ganglia are small but otherwise call for no special remark. The great splanchnic nerve is formed by the union of roots derived from thoracic ganglia 9, 10, and 11 compared with Zuckerman's (1938) finding in two specimens of *Hapale jacchus,* where it arose from the eighth ganglion with an accession of fibres from the twelfth ganglion. In one specimen he found a prominent splanchnic nerve from the seventh ganglion. Only the one splanchnic was encountered in *Callimico.*

White rami communicantes are given off as far caudally as the third lumbar segment, as Zuckerman found in *Hapale.*

From the last two lumbar segments roots are given off to constitute a hypogastric nerve which crosses the common iliac vessels ventrally to gain the pelvic viscera.

The main sympathetic chain is continued into the pelvis dorsal to the iliac vessels as a very fine nerve upon which distinct ganglia could not be seen, but a representative of the ganglion impar was found near the root of the tail.

Sacral autonomic fibres are given off only (or principally) from the first sacral nerve root. These are continued into a leash of fibres accompanying the pelvic vessels and constituting the nervi erigentes.

SENSE ORGANS

1. NASAL FOSSA

So far as could be observed without unduly damaging the skull, i.e., by inspection through the anterior nares after clearing away the nasal cartilages and through an opening chiselled through one half of the hard palate, the following data emerge.

There is a well-developed double-scrolled maxilloturbinal projecting medially some 1.5 mm., with an upward and a downward scroll, the former visible through the nares. Above this the lower ethmo-turbinal descends, its lower limit being visible through the nares.

From below the opening of the naso-lachrymal duct is visible beneath the anterior end of the maxilloturbinal.

There is a large maxillary antrum extending from the level of the canine back to the last molar. It measures 11.5 mm. long, 6 mm. broad and 4 mm. deep opposite P.².

2. EYEBALL

The sagittal diameter of the globe is 11.6 mm. and the transverse equatorial diameter 12.2 mm. compared with Menner's (1931) figures of 13.5 mm., and 13.5 mm. for *Hapale jacchus*. The pupil was fixed at a diameter of 1.8 mm., and the iris measures 2.2 mm. in depth.

The diameter of the cornea transversely is 6.55 mm. Circumcorneal pigmentary deposit affects a zone 0.4 mm. deep. The color of the iris registers between hazel and chestnut-brown on Ridgway's scale.

There is a nictitating membrane 1.5 mm. broad. The puncta lachrymalia are situated 1.5 mm. from the medial palpebral canthus.

3. AUDITORY ORGANS (fig. 77)

The external auditory meatus differs in no important respect from that in other platyrrhine monkeys as detailed by Joseph (1877). The cutaneous lining is provided throughout with short golden-brown hairs directed radially inwards and somewhat laterally. The diameter of the lumen near its fundus is 2.2 mm. The wall of the cartilaginous meatus immediately below the attachment of the tympanic membrane is 1.5 mm. thick, where it is thicker than elsewhere.

The tympanic membrane, 3.1 mm. in diameter, is completely circular and obliquely disposed, facing laterad, and slightly downwards and forwards. The umbo is situated slightly supero-anteriorly to the central point. The handle of the malleus and the chorda tympani nerve can be seen through the membrane.

The auditory ossicles resemble those of *Hapale* rather than of *Oedipomidas* or any of the Cebidae described and depicted by Doran (1878).

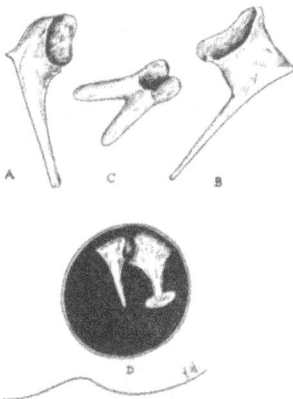

FIG. 77. A. Left malleus in lateral view. B. The same from the medial aspect. C. Left incus, medial view. D. Dissection of the left tympanic cavity with the ossicles *in situ*.

The malleus, 2.9 mm. long, has the same flattened capitulum as in *Hapale*, but the articular surface for the incus is relatively larger. There is the same spicular processus anterior and a slender manubrium terminating in a slight rounded knob.

The incus differs from that of *Hapale* in the lesser disproportion between the length of its two crura. In this it agrees with Doran's figure of the incus in *Oedipomidas*, but the two limbs are much longer and more divergent than in that genus. The articular surface is strongly grooved and, therefore, divided into upper and lower facets, as in *Hapale*, but the facets are more extensive on the medial surface of the body of the bone—in conformity with the larger articular area on the malleus.

The stapes is of the usual platyrrhine form, with straight, widely divergent crura, a relatively small foramen and a long, narrow foot-plate, twice as long as broad.

DISCUSSION

The currently recognized intermediate status of *Callimico* is based solely upon data from the external anatomy of the extremities combined with the dental formula, the former being hapalid and the latter cebid. So far these contrasting features have been made the basis for the erection of a distinct subfamily (Callimiconinae) variously assigned to the Hapalidae (= Callithricidae) or the Cebidae. Thus, Thomas, Weber (1924), and Simpson (1945) relegated the Callimiconinae to the Cebidae, considering the dental characters more significant than the nature of the cheiridia. Pocock (1925) and Wood Jones (1929) are exponents of the opposite view, as also, by implication from his views on the status of the Hapalidae, would be Gregory (1920). Dollman (1933), emphasizing its fully intermediate status, went further and demanded full family recognition (Callimiconidae), a solution accepted tentatively by the present writer (Hill, 1957; see, however, modified view expressed in the preface of that work).

The purpose of the present discussion will, therefore, be to analyze the evidence derived from the preceding account of the full anatomical details of *Callimico*, in favor of one or other of the above-mentioned alternative hypotheses and to ascertain thereby the place of *Callimico* in the course of Primate phylogeny, particularly in relation to that of other neotropical forms.

It must be emphasized that the data presented here are based mainly upon a single individual and, therefore, allowance cannot be made for the range of possible variation. However, the rarity of the species must be pleaded as sufficient justification for making the attempt to supply further information to enable us to supplement the meager data so far available from the museum taxonomists' standpoint. It should further be pointed out that the view that internal anatomical structures are more variable than external, often advocated by taxonomists, is quite untenable—a point recently stressed by Fisher and Goodman (1955) in their study of the myology of the Whooping Crane (*Grus americana*). These same authors admit that too often anatomists have been responsible for conclusions based upon the dissection of one or two specimens, which leaves their work open to criticism, but, nevertheless, in a situation like the present, where the species has been known to science for over half a century on the evidence of not more than limited parts of six individuals, the anatomist can do little less than report in the greatest detail possible his findings on such few individuals as may be available to him—particularly so where the information hitherto at hand indicates that the animal is aberrant taxonomically.

Throughout the foregoing account, details appear in which *Callimico* aligns itself basically with the Hapalidae, sometimes with the typical genus *Hapale*, but often with less well-known members of the family which, on the whole, are less specialized (e.g., as regards dentition) than *Hapale*, namely with *Tamarin*, *Tamarinus*, *Oedipomidas*, and *Leontocebus* (= *Leontideus*). As examples the following may be listed:

1. Microscopic characters of the hair.
2. Characters of the terminal phalanges and their cutaneous appendages.
3. Presence of glandular pustules on scrotum.
4. Dolichocephaly.
5. Horizontal plate of palate bone entirely within orbit.
6. Absence of cribriform plate on ethmoid, where a single foramen transmits one large olfactory nerve each side.
7. Vertebral formula (except for more numerous caudals).
8. Anticlinal vertebra is ninth thoracic.
9. Suprapatella present.
10. Absence of adductor tubercle on femur.
11. Mandible and masticating muscles.
12. Absence of occipital head of trapezius.
13. Lack of clavicular head of pectoralis major.
14. Independence of pyriformis from mesoglutaeus.
15. Morphology of stomach and duodenum.
16. Arrangement of taeniae coli.
17. The whole of the respiratory tract.
18. Low division of brachial artery.
19. Main features of brain (including cerebellum).

Many others could be listed, but the above are sufficiently numerous and topographically widespread to indicate general similarity with the Hapalidae as a group. Doubtless some are adaptive features attributable to similar environmental and ecological demands (e.g., site of anticlinal vertebra) but in the main these features show evidence of genetic community.

Some of the characters listed need further comment. The nails, for example, though agreeing in general conformation with those of other Hapalids, are less extreme in their clawlike character and approach the conditions met with in some of the more primitive cebids such as *Callicebus* and *Aotes*, and even such specialized animals as *Saimiri*.

The presence of glandular pustules on the scrotum is not confined to the Hapalidae, and even there is not uniform. In *Callimico* they are scattered and small as in *Oedipomidas* and *Leontocebus*, not heavily developed and congested as in *Hapale*, *Mico* and *Cebuella*. Sparse pustules occur also in some Cebidae, as for example in *Cebus* and *Saimiri*.

Dolichocephaly is shared more especially with *Leontocebus*, but also occurs, doubtless independently derived, in the cebid *Saimiri*.

In addition to the above, *Callimico* shares many characters with the Hapalidae which are primitive features inherited by both from earlier primate (? Tarsioid) ancestors. *Callimico*, however, further shows tarsioid or prosimian characters additional to those met with in the Hapalidae, e.g.:

1. Morphology of the penis, including absence of baculum, (shared by *Tarsius* and *Homo*).
2. Entepicondylar foramen on humerus.
3. Retention of coraco-brachialis longus.
4. Presence of sterno-clavicularis anterior.
5. Independence of pectoralis abdominis.
6. Constitution of triceps brachii.
7. Retention of four contrahentes in manus.
8. Generalized condition of deep manual extensors.
9. Simpler tensor fasciae femoris.
10. Dual pectineus.
11. Primitive mammalian arrangement of hamstrings.
12. Prosimian type of sartorius.
13. Flexor brevis hallucis with only one head.
14. Palatal rugae.
15. Short central section of mesocolon.
16. More primitive Spigelio-caudate complex in liver.
17. Muscular fibres in stroma of Cowper's gland.
18. Trilobed spleen (a marsupial feature).
19. Absence of oblique sinus of pericardium correlated with single opening of pulmonary veins into left atrium.
20. Sagittal situs of heart.
21. Arrangement of cusps at entrance of postcava unique but closer to prosimian than hapalid condition, where, however, *Oedipomidas* is nearest to *Callimico*.
22. Type C (prosimian) aortic arch (*Leontocebus* closest among Hapalidae).
23. Primitive arrangement of ileo-caecal vessels.
24. Single (left) inferior phrenic artery (as in *Tarsius*).

In addition to the dental formula, agreement with or approach toward the Cebidae occurs in the following particulars:

1. More vertical position of lower incisors (Dollman, 1937).
2. More distinct chin on mandible (Elliot, 1913).
3. Pigmentation of buccal mucosa.
4. Obsolescence of posterior palatine spine.

5. Caudo-femoralis attachments.
6. More complete differentiation of femoral adductors.
7. Greater development of tibialis posterior.
8. Absence of tendon of plantaris.
9. Certain details of the liver.
10. Presence of two perforating branches of profunda femoris (only one in Hapalidae; up to four in Cebidae).
11. Slight advances in cerebral anatomy (some, however, shared by *Oedipomidas*).
12. Minor advances in cerebellum wherein an approach is made toward *Callicebus* and *Aotes*.
13. Minor advances in cervical plexus.
14. Certain non-hapaline features in sacro-pudendal plexus.
15. Hypoglossal fila.
16. Minor details in auditory ossicles (shared again by *Oedipomidas*).

Finally, as is to be expected, *Callimico* presents a few features peculiar to itself, apparently specialized characters brought on by evolutionary divergence since its departure from the parental platyrrhine stem. These are:

1. Certain features in the external ear and in hair arrangement.
2. Ossification of the suprascapular ligament.
3. Aberrant tenuissimus muscle.
4. Loss of contrahens V in foot.
5. Absence of bloodless fold at ileo-colic junction.
6. Peculiar position of hilum of kidney.
7. Recurrent course of renal arteries.
8. Arrangement of cusps of Eustachian valve (see also, however, no. 21 in list 2).

It is not practicable to give a categorical statement on all the features in this last series. Some (e.g., 2, 3, 4) may be purely individual variations and are liable to crop up occasionally in any species of Primate. The characters of the external ear and the renal features are scarcely susceptible to this explanation and may therefore be peculiar to *Callimico*. On the whole the small number of these specific features is noteworthy. Taken in conjunction with (*a*) the large number of characters reminiscent of prosimian or even primitive mammalian anatomy, and (*b*) with the other primitive features shared with the Hapalidae it is manifest that *Callimico*, despite its approach towards certain cebids, is a very primitive pithecoid Primate. Its likenesses to the Hapalidae are, however, important and within that group it clearly shares more in common with such genera as *Tamarin*, *Oedipomidas*, and *Leontocebus* than with the somewhat more specialized *Hapale*, *Mico*, and *Cebuella*

Of the cebid features, which are relatively few, the approach is for the most part towards the more primi-

tive genera, particularly *Callicebus* and *Aotes*, and is shared by *Callimico* in lesser degree with the more primitive genera of the Hapalidae. It is in the dentition that the most salient cebid features are found, not only in the retention of the third molars but also, as shown by Dollman (1937), in the more vertically implanted lower incisors and in the crown characters of the lower molars.

Germane to this question, therefore, is Gregory's discussion on the status of the Hapalidae, based on his study of the cranial skeleton and, more particularly, the dentition. Gregory concluded, and in this he was followed by Pocock, that despite the retention of many primitive features the Hapalidae are not the most generalized of pithecoid Primates as is commonly believed (see especially Bolk, 1916)—a position still accepted by Piveteau (1957). On the contrary they are a specialized offshoot from some primitive platyrrhine stock so far unrepresented in the fossil record. The Hapalidae are, in fact, less primitive than some of the Cebidae. Their crania can be derived from the ancestors of *Callicebus* and *Aotes*, from which they differ in the enfeeblement of the lower jaw and the zygomatic arch and the loss of the third molars in both upper and lower jaws; in the reduced size of the orbits, enlargement of canines and antero-posterior elongation of the brain-case. There is no necessity here to repeat all Gregory's arguments, but it seems that *Callimico* amply supports his arguments in so far as it fills a gap in his suggested phylogenetic series.

Callimico, in its dental formula, represents a stage in the evolution of the Hapalidae prior to the final loss of the last molars—a stage incipient already in some Cebidae, especially *Callicebus, Aotes,* and *Saimiri*.[1] Its close relationship with the Hapalidae in other details of its anatomy, together with its retention of many prosimian and tarsioid features not present in the majority of Hapalidae, places it in a basal position on the stem of the New World Primates alongside *Callicebus* in particular for, as Gregory declares, *Callicebus* is "the most tarsioid New World monkey" in contrast to such forms as *Cebus* and *Saimiri* which are, each in its own way, the most advanced and pithecoid (this last statement of Gregory's, however, is open to some objection, for in most respects the Atelinae are more specialized still, though in yet other directions).

The adjoining scheme is, therefore, presented as representing the probable relationships of the New World Primates here considered according to the evidence at present available:

---

[1] In this connection too Colyer (1936) reports the occasional absence of M_3^1 in *Chiropotes albinasa*, while Stirton (1951) states that in the Miocene *Cebupitherfia* M_3^3 are reduced.

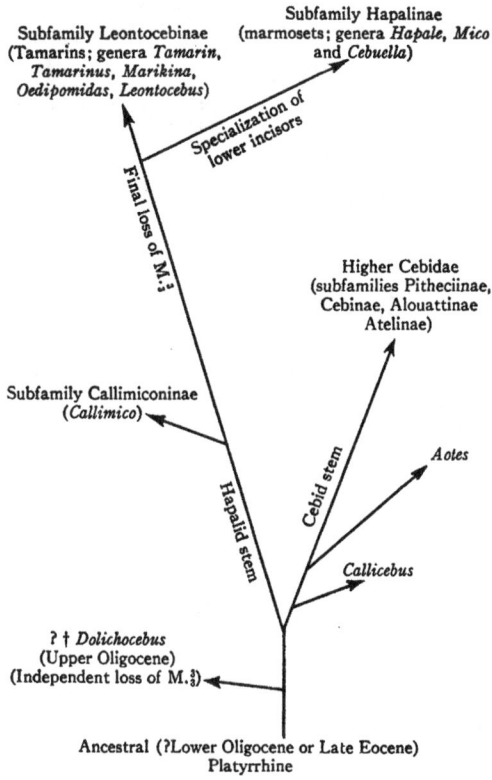

REFERENCES TO LITERATURE

APPLETON, A. B. 1928. The muscles and nerves of the postaxial region of the tetrapod thigh. *Journal Anat. Lond.* **62**: 364–400, 401–438.

AYER, A. A. 1940. The origin of the pisiform. *Curr. Sci. Bangalore* **9**: 333.

——. 1948. The anatomy of Semnopithecus entellus. Madras, Indian Publishing House.

BANG, F. B. 1936. Observations on the limb arteries of the Woolly monkey (*Lagothrix lagothrica*). *Anat. Rec.* **66**: 387–395.

BAYER, L. 1892. Beitrag zur vergleichenden Anatomie der Oberarmarterien. *Morph. Jb.* **19**: 1–41.

BEATTIE, J. 1927a. The visceral lymphatic channels of the Catarrhinae (illustrated by the direction of the spread of tuberculosis). *Proc. Zool. Soc. Lond.* 1927 (1):459–470.

——. 1927b. The anatomy of the Common Marmoset (*Hapale jacchus* Kuhl). *Proc. Zool. Soc. Lond.* 1927 (2): 593–718.

BEDDARD, F. E. 1907. On the azygos veins in the Mammalia. *Proc. Zool. Soc. Lond.* 1907: 181–223.

VON BISCHOFF, T. L. W. 1870. Beiträge zur Anatomie des *Hylobates leuciscus* und zur einer vergleichenden Anatomie der Muskeln der Affen und des Menschen. *Abh. Bayer. Akad. Wiss.* **10**, abt. 3: 197–297.

BOLK, L. 1902a. Beiträge zur Affen-Anatomie, III, Der Plexus cervicobrachialis der Primaten. *Ned. Bijd. Anat.* **1**: 371–566.

——. 1902b. Beiträge zur Affen-Anatomie IV, Das Kleinhirn der Neuweltaffen. *Morph. Jb.* **31**: 44–84.

——. 1916. Problems of human dentition. *Amer. Jour. Anat.* **19**: 91–148.

BRADLEY, O. C. 1904–1905. The mammalian cerebellum, its lobes and fissures. *Jour. Anat. Lond.* **38**: 448–475; **39**: 99–117.

BRUHNS, FANNY. 1910. Der Nagel der Halbaffen und Affen. Ein Beiträg zur Phylogenie des menschlichen Nagels. *Morph. Jb.* **40**: 501–609.

BRYCE, T. H. 1923. *Quain's Anat.* **4** (2): 102.

BURMEISTER, H. 1846. Beiträge zur Kenntnis der Gattung Tarsius. Berlin, G. Reimer.

CABRERA, A. 1922. Manual de Mastozoologie. Madrid and Barcelona.

——. 1956. Sobre la identificación de *Simia leonina* Humboldt (Mammalia, Primates). *Neotropica* **2**: 49–53.

CLARK, W. E. LE GROS. 1926. The mammalian oculo-motor nucleus. *Jour. Anat. Lond.* **60**: 426–448.

——. 1936. The problem of the claw in Primates. *Proc. Zool. Soc. Lond.* 1936: 1–24.

COLYER, F. 1936. Variations and diseases of the teeth of animals. London, Bale Sons & Danielsson.

CONNOLLY, C. J. 1950. External morphology of the primate brain. Springfield, C. C. Thomas.

CORDIER P., P. COULOUMA, DEVOS, and DELCROIX. 1936. Contribution à l'étude de la constitution du plexus cervical chez l'homme et quelques primates. *C. R. Ass. Anat.* **31**: 114–123.

CORDIER P., P. COULOUMA, and VAN VARSEVELD. 1936. Contribution à l'étude de la constitution du plexus sacré chez l'homme et quelques primates. *C. R. Ass. Anat.* **31**: 124–132.

CUNNINGHAM, D. J. 1882. Report on some points in the anatomy of the thylacine (*Thylacinus cynocephalus*), cuscus (*Phalangista maculata*) and phascogale (*Phascogale calura*), collected during the voyage of *H.M.S. Challenger* in the years 1873–1876: with an account of the comparative anatomy of the intrinsic muscles and nerves of the mammalian pes. *Challenger Reports* **5** (Zool.) (16): 192.

DAVIES, D. V. 1947. The cardio-vascular system of the slow loris (*Nycticebus tardigradus malaianus*). *Proc. Zool. Soc. Lond.* **117**: 377–410.

DAVIES, D. V., and W. C. O. HILL. 1954. The abdominal portion of the alimentary system in *Hapalemur* and *Lepilemur*. *Proc. Roy. Soc. Edinb. B.* **65**: 182–204.

DOLLMAN, G. 1933. Primates, series 3, British Museum (Nat. Hist.) booklet (no pagination).

——. 1937. Exhibition of skins of marmosets and tamarins. *Abstr. Proc. Zool. Soc. Lond.*, 64–65.

DORAN, A. H. G. 1878. Morphology of the mammalian ossicula auditus. *Trans. Linn. Soc. Lond.* (2), **1**: 371–497.

ECKSTEIN, F. M. P. 1944. The pisiform bone. *Nature*, London, **154**: 182.

ELLIOT, D. G. 1913. Review of the primates (Mon. Amer. Mus. Nat. Hist., no. 1) **1**: 224; **3** (appendix 2): 261–262.

ELLIOT SMITH, G. 1902. Descriptive and illustrated catalogue of the physiological series of comparative anatomy contained in the Museum of the Royal College of Surgeons of England 2 (2 ed.). London.

FICALBI, E. 1889. Contribuzione alla conoscenza della angeologia delle Scimmie. *Atti. Acad. Fisiocr. Siena*, 1889, (4), **1**: 425–456.

FIEANDT, E. 1914. Ueber das Wurzelgebiet des Nervus hypoglossus und den Plexus hypoglosso-cervicalis bei den Säugetieren. *Morph. Jb.* **48**: 513–643.

FISHER, H. I., and D. C. GOODMAN. 1955. The myology of the Whooping-crane, *Grus. americana. Illinois Biol. Monog.* **24** (2). Urbana, Illinois.

FÖRSTER, A. 1916. Der m. extensor tarsi (Peronaeus tertius?) bei *Hapale jacchus. Morph. Jb.* **52**: 257–276.

——. 1922a. La tuberosité du scaphoïde et le jambier posterieur. *Arch. Anat. Strassb.* **1**: 1–55.

——. 1922b. Étude de l'evolution phylogénétique de la vessie dans la série des mammifères superieurs; prosimiens et primates. *Arch. Anat. Strasb.* **1**: 205–244.

FRANSEN, J. W. P. 1907. Le système vasculaire abdominal et pelvien des Primates. Anatomie descriptive et relations segmentaires. *Ned Bijdr. Anat.* **4**: 125–183, 487–537.

GOELDI, E. *in* GOELDI, E., and G. HAGMANN. 1904. Prodromo de um catalogo critico, commentado da colleccão de mammiferos no Museu do Pará. *Bol. Mus. Goeldi* **4** (1): 38–122 (footnote, p. 54).

GREEN, H. L. H. H. 1931. The occurrence of a tenuissimus in a human adult. *Jour. Anat. Lond.* **65**: 266–271.

GREGORY, W. K. 1920. The origin and evolution of the human dentition, Pt. III. *Jour. Dent. Res.* **2**: 357–426.

——. 1922. Origin and evolution of the human dentition. Baltimore, Williams & Wilkins Co.

HANSTRÖM, B. 1948. A comparative study of the pituitary in the monkeys, apes and man. *K. Fysiogr. Sallsk. Handl.* N. F. **59** (10).

——. 1952. The pituitary of the marmosets. *Lunds Univ. Årsskr.* N. F., Afd. 2, **48**: 3–12.

——. 1953. The hypophysis in a wallaby, two tree-shrews, a marmoset and an orang-utan. *Ark. Zool. Stockh.* **6**: 97–154.

HARRIS, H. A. 1944. The pisiform bone. *Nature*, Lond., **153**: 715; **154**: 183.

HECKER, P. 1922. Appareil ligamenteux occipito-altoido-axoidien. *Arch. Anat. Strasb.* **1**: 413–436.

HERSHKOVITZ, P. 1957. The systematic position of the marmoset, *Simia leonina* Humboldt (Primates). *Proc. Biol. Soc. Washington* **70**: 17–20.

HILL, W. C. OSMAN. 1953. The blood vascular system of *Tarsius. Proc. Zool. Soc. Lond.* **123**: 655–694.

——. 1955. Primates II, Haplorhini; Tarsioidea. Edinb. Univ. Press.

——. 1956. *Callimico goeldii*, international approach to anatomy. *X-ray Focus* **1**: 15–16.

——. 1957. Primates III Pithecoidea. Edinb. Univ. Press.

——. 1958. *In* Karger, Handbuch der Primatologie, **3** (1). Basel.

HILL, W. C. OSMAN, H. M. APPLEYARD, and L. AUBER. 1959. The specialized area of skin in *Aotes* Humboldt (Simiae, Platyrrhini). *Trans. Roy. Soc. Edinb.* **63**: 535–551.

HILL, W. C. OSMAN, and D. V. DAVIES. 1954. The reproductive organs in *Hapalemur* and *Lepilemur*. *Proc. Roy. Soc. Edinb.* (B) 65: 251–270.

HILL, W. C. OSMAN, and D. V. DAVIES. 1956. The heart and great vessels in the Strepsirhini. *Trans. Roy. Soc. Edinb.* 63: 115–127.

HILL, W. C. OSMAN, and R. E. REWELL. 1948. The caecum of primates, its appendages, mesenteries, and blood-supply. *Trans. Zool. Soc. Lond.* 26: 199–256.

HOWELL, A. B. 1938. Morphogenesis of the architecture of hip and thigh. *Jour. Morph.* 62: 177–218.

HRDLIČKA, A. 1924. Brain and organ weights in American monkeys. *Amer. Jour. Phys. Anthrop.* 8: 201–211.

HÜBER, E. 1931. Evolution of facial musculature and facial expression. Baltimore, Johns Hopkins Press.

HUNTINGTON, T. S. 1904. The derivation and significance of certain supernumerary muscles of the pectoral region. *Jour. Anat. Lond.* 39: 1–54.

JAMIESON, E. B. 1904. The gluteal and femoral muscles, with their nerve supply, in a marmoset (*Hapale jacchus*). *Proc. Roy. Phys. Soc. Edinb.* 15: 168–194.

JOHNSTON, T. B. 1920. The ileo-caecal region of *Callicebus personatus*. *Jour. Anat. Lond.* 54: 66–78.

JOSEPH, G. 1874. Ueber das Verhalten des ausseren Gehörgangs und der Paukenhohle bei den amerikanischen Affen. *Jber. Schles. Ges. vaterl. Cult.*, 44–47.

VAN KAMPEN, P. N. 1905. Dei Tympanalgegend des Säugethierschadels. *Morph. Jb.* 34: 321–722.

KEITH, A. 1895. The modes of origin of the carotid and subclavian arteries from the arch of the aorta in some of the higher Primates. *Jour. Anat. Lond.* 29: 453–458.

KENNARD, M. A., and M. D. WILLSER. 1941. Weights of brains and organs of 132 New and Old World monkeys. *Endocrinology* 28: 977–984.

KOHLBRUGGE, J. H. F. 1897. Muskeln und periphere Nerven der Primaten, mit besonderer Berucksichtigung ihrer anomalien. *Verhandl. Akad. Wetens.* Amsterdam (2 sect.) 5: 1–246.

KOLLMANN, J. 1894. Der levator ani und der coccygeus bei den geschwantzen Affen und den Anthropoiden. *Anat. Anz.* 9: 198–205.

KOSINSKI, C. 1927. L'innervation cutanée de la face dorsale de la main, basée sur l'examen de 300 pieces anatomiques, avec quelques notions d'anatomie comparée. *C. R. Asso. Anat.* 22: 121–133.

LAMPERT, H. 1926. Zur Kenntnis des Platyrrhinenkehlkopfes. *Morph. Jb.* 55: 607–654.

LIGHTOLLER, G. S. 1934. The facial musculature of some lesser primates and a *Tupaia*. *Proc. Zool. Soc. Lond.* 1934 (1): 259–309.

LIMA, E. DA C. 1945. Mammals of Amazonia, 1, General introduction and primates. Belem do Pará and Rio de Janeiro, Mus. Paraense.

LINEBACK, P. 1933. *In* Hartman and Straus, The anatomy of the Rhesus monkey (*Macaca mulatta*). Baltimore, Williams and Wilkins.

MC.C...., R. M. 1955. We have a *Callimico*, but is it a marmoset or a monkey? *Anim. Kingd.* 58: 20–30.

MANNERS-SMITH, T. 1910–1912. The limb arteries of primates. *Jour. Anat. Lond.* 44: 271–302; 45: 23–64; 46: 95–172.

MENNER, E. 1931. Ueber die Retina einiger Kleinaffen aus den Familien Callithricidae und Cebidae. *Zool. Anz.* 95: 1–12.

MENSA, A. 1913. Arterie meningee encefaliche nella serie dei mammiferi. Studio morfoligico e descrittivo. *Morph. Jb.* 46: 1–207.

MILLER, RUTH. 1947. The inguinal canal of primates. *Amer. Jour. Anat.* 80: 117–142.

MÜLLER, E. 1904. Beitrage zur Morphologie des Gefasssystems. I. Die Armarterien des Menschen; II. Die Armarterien der Saugetiere. *Arb. anat. Inst. Wiesbaden = Anat. Heft.* 27: 71–242.

MURIE, J., and St. G. MIVART. 1872. On the anatomy of the Lemuroidea. *Trans. Zool. Soc. Lond* 7: 1–113.

OTTLEY, W. 1879. On the attachment of the eye-muscles in mammals. I. Quadrumana. *Proc. Zool. Soc. Lond.* 1879: 121–128.

OUDEMANS, J. T. 1892. Die accessorischen Geschlectsdrüsen der Säugethiere. Harlem.

PARKER, W. K. 1868. A monograph on the structure and development of the shoulder-girdle and sternum in the Vertebrata. Ray Soc. monogr. Lond.

PIVETEAU, J. 1957. Traité de Paleontologie 7, Primates. Paris, Masson.

POCOCK, R I. 1917. The external characters of the Hapalidae. *Ann. Mag. Nat. Hist.* (8) 20: 247–258.

——. 1920. On the external characters of the South American monkeys. *Proc. Zool. Soc. Lond.*, 91–113.

——. 1925. Additional notes on the external characters of some platyrrhine monkeys. *Proc. Zool. Soc. Lond.*, 27–47.

POPOWSKI, J. P. 1894. Das Arteriensystem der unteren Extremitäten bei den Primaten. *Anat. Anz.* 10: 55–80, 99–114. See also 1893, *Anat. Anz.* 8: 657–665.

——. 1895. Arterialnaya sistema u obezyan sravnitelno s raspolzheniyem yaya u chlovieka. *Isv. Imp. Tomsk Univ.* 8: 1–152.

RANKE, K. 1897. Muskel und Nervenvariationen der dorsalen Elemente des Plexus ischiadicus der Primaten. *Arch. Anthropol.* 24: 117–144.

RAU, A. S., and P. K. RAO. 1930. The arterial system of *Loris lydekkerianus*. *Jour. Mysore Univ.* 4: 90–121.

RETTERER, E., and H. VALLOIS. 1912. De la double rotule de quelques primates. *C. R. Soc. Biol.* 73: 379–382.

RIBEIRO, A. de Miranda. 1912. Zwei neue Affen unserer Fauna (Dois novos simios da nossa fauna). *Brasilianische Rundschau* 2, 21–23. Reprinted 1955 in *Arqu. Mus. Nac. Rio de Janeiro* 42 (i): xxxix-xli.

——. 1940. Commentaries on South American primates. *Mem. Inst. Oswaldo Cruz* 35: 779–851.

ROJECKI, F. 1889. Sur la circulation arterielle chez le Macacus. *Jour. Anat. Paris* 25: 343–386.

RUGE, G. 1878. Zur vergleichenden Anatomie der tiefen Muskeln in der Fusssohle. *Morph. Jb.* 4: 644–659.

——. 1887. Untersuchungen uber die Gesichtsmuskulatur der Primaten. Leipzig, W. Engelmann.

——. 1893. Verschiebungen in den Endgebiet der Nerven des Plexus lumbalis des Primaten. *Morph. Jb.* 20: 305–397.

——. 1902. Die ausseren Formverhaltnisse der Leber bei den Primaten III. Die Leber der platyrrhinen Westaffen. *Morph. Jb.* 30: 42–84.

SANDERSON, I. T. 1940. Mammals of the North Cameroons forest area. *Trans. Zool. Soc. Lond.* 24: 623–725.

SCHREIBER, H. 1928. Die Gesichtsmuskulatur der Platyrrhinen. *Morph. Jb.* 60: 179–295.

SCHULTZ, A. H. 1921. Fetuses of the Guiana howling monkey. *Zoologica N. Y.* 3: 243–262.

——. 1949. The palatine ridges of primates. *Contr. Embryol. Carneg. Instr.*, Washington, 33: 43–66.

SCHULTZ, A. H., and W. L. STRAUS. 1945. The numbers of vertebrae in primates. *Proc. Amer. Philos. Soc.* 89: 601–626.

SHUFELDT, R. W. 1914. On the osteology of the genera *Lasiopyga* and *Callithrix*, with notes upon the osteology of the genera *Seniocebus* and *Aotus*. *Ann. Carneg. Mus.* 9: 58–85.

SIEGLBAUR, F. 1931. Os marginale manus ulnare. *Wien Klin. Wschr.* 44: 832–838.

SILVESTER, C. F. 1912. On the presence of permanent communications between the lymphatic and the venous system at the level of the renal veins in adult South American monkeys. *Amer. Jour. Anat.* 12: 447–471.

SIMPSON, G. G. 1945. The principles of classification and a classification of mammals. *Bull. Amer. Mus. Nat. Hist.* 85: 1–350.

SONNTAG, C. F. 1921. The comparative anatomy of the tongues of the Mammalia IV. Families 3 and 4, Cebidae and Hapalidae. *Proc. Zool. Soc. Lond.*, 497–524.

SSOKOLOW, P. 1933. Zur Anatomie des N. suralis beim Menschen und Affen. *Z. Anat. Entw. Gesch.* 100: 194–217.

STARCK, D. 1933. Die Kaumuskulatur der Platyrrhinen. *Morph. Jb.* 72: 212–285.

STERZI, A. I. 1903. Richerche sopra le anastomosi dei rami anteriori del plesso brachiale e loro interpretazione morfologica. *Arch. Ital. Anat. Embriol.* 2: 178–205.

STIRTON, R. A. 1951. Ceboid monkeys from the Miocene of Colombia, *University of Calif. Publ. Bull. Dept. of Geol.* 28: 315–356.

STRAUS, W. L. 1930. The foot musculature of the highland gorilla. *Quart. Rev. Biol.* 5: 261–317.

TANDLER, J. 1899. Zur vergleichenden Anatomie der Kopfarterien bei den Mammalia. *Denkschr. Akad. Wiss. Wien* 67: 677–784.

THEILE, W. 1852. Uber das Arteriensystem von *Simia inuus*. *Arch. anat. Physiol. Wiss. Med.*, 419–449.

THOMAS, O. 1904. New *Callithrix, Midas, Felis, Rhipidomys* and *Proechimys* from Brazil and Ecuador. *Ann. Mag. Nat. Hist.* (7) 14: 188–196.

——. 1913. Exhibition of a specimen of the Amazonian monkey, *Callimico snethlageri*. *Proc. Zool. Soc. Lond.*, 3–4.

——. 1913. On some rare Amazonian mammals from the collection. *Ann. Mag. Nat. Hist.* (8) 11: 130–136.

——. 1914. On various South American mammals. *Ann. Mag. Nat. Hist.* (8) 13: 345–363.

——. 1928. The Godman-Thomas Expedition to Peru. VII. The Mammals of the Ucayali. *Ann. Mag. Nat. Hist.* (10) 2: 249–265.

WATERMAN, H. C. 1929. Studies on the evolution of the pelvis of man and other primates. *Bull. Amer. Mus. Nat. Hist.* 58: 585–642.

WEBER, M. 1924. Die Säugetiere. 2nd ed. Jena, G. Fischer.

WINCKLER, G. 1930. La branche thénarienne du nerf median. *Arch. Anat. Strasbourg* 12: 151–227.

——. 1934. Le nerf péronier accessoire profond. *Arch. Anat. Strassb.* 18: 183–219.

WINDLE, B. C. A. 1885–1886. Notes on the myology of *Midas rosalia*, with remarks on the muscular system of apes. *Proc. Bgham. Nat. Hist. Soc.* 5: 152–166.

——. 1886–1887. Notes on the myology of *Hapale jacchus*. *Proc. Bgham Nat. Hist. Soc.* 5: 277–281.

WOOD JONES, F. 1929. Man's place among the mammals. London, Arnold.

——. 1940. The nature of the soft palate. *Jour. Anat. Lond.* 74: 147–170.

WOOLLARD, H. H. 1925. The anatomy of *Tarsius spectrum*. *Proc. Zool. Soc. Lond.*, 1071–1184.

——. 1926. Notes on the retina and lateral geniculate body in *Tupaia, Tarsius, Nycticebus* and *Hapale*. *Brain* 49: 77–105.

ZUCKERKANDL, E. 1898. Zur Anatomie von *Chiromys madagascariensis*. *Denkschr. Akad. Wiss. Wien* 68: 89–200.

——. 1900. Zur Morphologie der Arteria pudenda interna. *S. B. Akad. Wiss. Wien*, abt. 3, 109: 405–458.

ZUCKERMAN, S. 1938. On the automatic nervous system and on vertebral and neural segmentation in monkeys. *Trans. Zool. Soc. Lond.* 23: 315–378.

INDEX

Acetabulum, 30, 35
Air-cells (petro-tympanic), 21, 22
Air sinuses (paranasal), 20, 23
Alouatta, 71, 73, 85
Antrum, maxillary, 23, 107
Aotes, 19, 68, 70, 72, 92, 98, 109, 110
Arbor vitae, 98
Arctocebus, 17, 85
Ateles, 60, 73, 85, 93, 94, 106

Balearica regulorum, 10
Brown fat, 37, 43
Bulla, tympanic, 19, 20, 21, 22, 41, 43, 45, 88, 90, 100
Bursa, infracardiac, 77, 82
Bursae, 31, 34, 35, 57

Cacajao, 70, 72
Calcaneus, 31, 60, 61, 63
Callicebus, 9, 68, 70, 72, 80, 96, 104, 109, 110
Callimico snethlageri, 9, 10
Callithrix, 9, 104
Canal, inguinal, 46
Cebuella, 10, 23, 24, 65, 66, 69, 70, 71, 85, 109, 110
Cebupithecia, 110
Cebus, 11, 40, 56, 66, 71, 85, 100, 101, 103, 104, 105, 106, 110
Cebus apella, 73, 85
Cebus capucinus, 73, 107
Cebus unicolor, 71, 73
Cebus xanthosternos, 71, 72, 85
Cercocebus, 33
Cercopithecus, 33
Chevron bones, 26
Chiropotes, 10, 110
Circle of Willis, 90
Cnemial crest, 30
Contrahentes, 53, 61, 103, 107, 110
Cyamella, 36

Daubentonia, 61
Dermatoglyphics, 15

Ear, 10, 13, 108
Epiglottis, 64, 73, 74, 75
Eye, 13, 38, 39, 98, 108
Eyelids, 11, 13, 90

Fabellae, 35, 59
Fold, aryteno-epiglottidean, 73, 75; glosso-epiglottic, 73; palato-glossal, 45, 67; palato-pharyngeal, 45, 67; rectoduodenal, 68; vestibular, 73; vocal, false, 73; true, 73
Frenal lamella, 11, 64, 65

Galago, 85
Gall-bladder, 71
Ganglion, ciliary, 98; Gasserian, 21, 99; impar, 107; otic, 99; stellate, 102, 103, 107; superior cervical, 107; trunci vagi, 100, 101

Gland, bulbo-urethral (Cowper's), 79; circumanal, 18; cutaneous, 18; hibernating, 37; mammary, 18; parotid, 66, 89, 100; prostate, 78, 79, 93; sublingual, 66; submandibular, 66
Glandular fat, 37, 43
Grus americanus, 109

Hair, 10, 11, 14; micro-anatomy of, 11, 12
Hallux, 11, 15, 32, 61, 62, 63
Hapale, 10, 12, 17, 21, 22, 23, 24, 25, 27, 28, 29, 30, 31, 32, 35, 37, 38, 39, 40, 41, 42, 43, 44, 45, 47, 48, 49, 50, 52, 54, 55, 56, 57, 58, 59, 60, 61, 63 64, 65, 66, 69, 74, 75, 77, 81, 82, 85, 90, 92, 94, 95, 96, 98, 100, 101, 102, 103, 105, 106 109, 110
Hapale humeralifer, 96
Hapale jacchus, 12, 24, 48, 69, 70, 71, 73, 84, 86, 107, 108
Hapale penicillata, 37, 40, 67, 84, 89, 93, 96, 97
Hapalemur, 69, 79, 83, 85
Homo, 90, 103, 109

Interosseous muscles, 53, 62, 103, 107
Iris, 13

Lagothrix, 71, 73, 80, 94
Latissimo-epicondyloideus, 48, 51
Lemur, 61, 69, 79, 85
Lemurs, 28, 58, 67, 97
Leontideus, 9, 10, 15, 20, 21, 24, 101, 109
Leontocebus, 9, 10, 15, 20, 21, 24, 41, 48, 50, 54, 56, 59, 60, 61, 64, 65, 66, 69, 70, 71, 81, 82, 96, 101, 102, 103, 109, 110
Leontocebus rosalia, 22, 40, 70, 72, 84, 85
Leontocebus tamarin, 9
Ligament or Ligamentum, bifurcate, 36; check, 33; coraco-clavicular, 27; coraco-humeral, 34; coraco-scapular, 27, 28, 42; coronary, of knee, 35, 36; coronary, of liver, 72; cruciate, 36; dorsal calcaneo-cuboid, 36; falciform, of liver, 70, 71; hepato-cavo-phrenicum, 72, 73; inguinal, 45; interclavicular, 33; interspinous, 48; L. latum (pulmonis), 76; L. latum (uteri), 80; nuchal, 42; oblique (of spine), 33; occipito-odontoid, 33; orbicular, 34, 55; ovarian, 80; patellar, 59; pisi-hamate, 29; pisi-metacarpal, 29; plantar calcaneo-cuboid, 36; plantar intercuneiform, 37; spring, 31, 36, 37; stylo-mandibular, 41; tarsal interosseous, 31, 36, 37; teres (of hip), 35; thyro-hyoid, 74; transverse (of atlas), 33; transverse humeral, 34; transverse intermetacarpal, 29
Linea alba, 45
Lorises, 67
Lorisoidea, 83
Lumbrical muscles, 52, 53, 61, 103, 107

Macaca, 46, 89
Man, 30, 31, 35, 36, 37, 45, 48, 49, 58, 68, 69, 80, 82, 89, 90, 91, 92, 94, 95, 99
Mandrillus, 89
Manus, 11, 15
Marikina, 10, 13
Marmoset, 11, 15, 19, 30, 67, 69
Meatus auditorius externus, 20, 21, 100
Meatus auditorius internus, 21, 99, 100
Meatus urinarius, 17, 78
Mediastinum, 82, 86, 91
Membrana tectoria, 33
Membrane, atlanto-occipital, 33, 92; crico-thyroid, 74, 75; crico-vocal, 74; interosseous (of forearm), 34, 55, 93; interosseous (of leg), 60, 62; nictitating, 11, 13, 108; obturator, 57, 58, 94; thyro-hyoid, 73
Mesocyst, 77
Mesohepar, 72
Mico, 9, 10, 24, 66, 85, 95, 109, 110
Midas goeldii, 9
Midas ursulus, 9
Muscle, tensor tympani, 21

Nails, 1, 15
Nervi erigentes, 107
Nycticebus, 81, 91

Oedipomidas, 10, 13, 24, 37, 45, 64, 66, 69, 70, 71, 72, 84, 96, 97, 108, 109, 110
Orbit, 19, 22, 23, 39, 40, 90, 98, 99, 110
Os capitatum, 29
Os centrale, 29
Os cuboides, 32
Os hamatum, 29
Os lunatum, 29
Os magnum, 29
Os naviculare, 31, 60
Os pisiforme, 29
Os scaphoides, 29
Os trapezium, 29, 53
Os trapezoides, 29
Os triquetrum, 29
Ossa cuneiformes, 22, 60, 62

Papillae, lingual, 64, 65; vallate, 64, 65; of cervix, 81; of vagina, 81
Pelage, 10
Penis, 15, 17
Perodicticus, 17, 85
Pigment, cutaneous, 11
Pithecia, 70, 72
Pollex, 15, 52, 53, 55
Pongidae, 103
Propithecus, 83

Rhinarium, 11, 15

Saimiri, 11, 19, 70, 72, 94, 96, 104, 109, 110
Sclera, 11, 13
Scrotum, 17, 18, 78
Septum, nasale, 13, 23; presphenoidale, 23, 24

115

www.ingramcontent.com/pod-product-compliance
Lightning Source LLC
Chambersburg PA
CBHW081337190326
41458CB00018B/6029